Construction Equipment Policy

McGRAW-HILL SERIES IN CONSTRUCTION
ENGINEERING AND MANAGEMENT
Clarkson H. Oglesby, Consulting Editor

Douglas: Construction Equipment Policy
Parker and Olgesby: Methods Improvement for Construction Managers

CONSTRUCTION EQUIPMENT POLICY

James Douglas
Stanford University

McGraw-Hill Book Company
New York   St. Louis   San Francisco   Auckland   Düsseldorf
Johannesburg   Kuala Lumpur   London   Mexico   Montreal   New Delhi
Panama   Paris   São Paulo   Singapore   Sydney   Tokyo   Toronto

Library of Congress Cataloging in Publication Data

Douglas, James, date
    Construction equipment policy.

    (McGraw-Hill series in construction engineering and management)
    1. Construction equipment.   2. Construction industry—Management.   3. Engineering economy.   I. Title.
TH900.D68   658.2′7   74-14693
ISBN 0-07-017658-2

CONSTRUCTION EQUIPMENT POLICY

Copyright © 1975 by McGraw-Hill, Inc. All rights reserved. Printed in the United States of America. No part of this publication may be reproduced, stored in a retrieval system, or transmitted, in any form or by any means, electronic, mechanical, photocopying, recording, or otherwise, without the prior written permission of the publisher.

1 2 3 4 5 6 7 8 9 0 KPKP 7 9 8 7 6 5

This book was set in Times Roman by Textbook Services, Inc.
The editors were B. J. Clark and J. W. Maisel;
the production supervisor was Charles Hess.
The drawings were done by Eric G. Hieber Associates Inc.
Kingsport Press, Inc., was printer and binder.

CONTENTS

		Preface	vii
Chapter	1	Introduction to Equipment Policy	1
	2	Equipment Economics	15
	3	Some Economic Influences	28
	4	Life of Equipment	47
	5	Solution for Equipment Life	74
	6	Cost, Time, and Production Records	96
	7	Equipment Financing	117
	8	Standardization	151
	9	Inventory Management	176
	10	Maintenance Management	192
	11	Equipment Safety	220
Appendix	1	List of Variables	242
	2	Summary of Equations	245
	3	Table of Factors	248
	4	Chart of Accounts	289
	5	Equipment Classification Code	294
		Index	299

PREFACE

This book has been written for many reasons, but the principal one is to fill a void which now exists in the literature of construction management. It will be found useful by students of construction management whether they are still in school or in the hierarchy of executive management. For those managers in the construction industry, it will augment the mid-career education now so necessary to keep up with new technology.

This book is not highly technical; it can be easily understood by the intelligent observer with or without a college degree. Too, it is the result of many years' experience and research in the subject of construction-equipment policy by an author who fortunately had both field training and management experience.

After returning to the educational environment subsequent to a number of years of active engagement in construction, the author developed a course in this subject which is presently being taught at Stanford University in the program in construction management. Several years of refinement by bright students have helped to expand knowledge of the subject. Their critical observations have helped to keep the subject matter at an understandable level.

In preparing this book, the principal aspects of equipment policy were sorted out and each chapter relates to a subject or combination of them. Chapter 1 is a general coverage of the subject of equipment policy; it tells what should be included in policy and why. Some history of the technological progress which has taken place in the past century serves to emphasize why the owner of equipment needs to organize his thinking about equipment policy and utilize some of the recent sophisticated methods of decision analysis in formulating this policy.

Chapter 2 discusses engineering economy and its application to equipment. Consideration of the differences between alternatives, the time value of money, sunk costs, and other well-known principles of engineering economy are illustrated. Examples of these techniques related to equipment are given with simple numbers so they can be understood easily by the average reader.

The most important economic factors which influence equipment economics are covered in Chap. 3. These include the available methods of depreciation, salvage value, and tax effects such as gain on sale, investment credit, and recapture of income. Inflation is defined and illustrated in simple terms which the layman can understand. Lastly, the effects of technological progress through obsolescence are described and are related to the economics of equipment.

In Chap. 4 the basic economic theories of equipment life and replacement are discussed. Simple examples illustrate minimizing costs and maximizing profits. The development of a mathematical model for the solution of economic life and replacement life is described. The equations developed for this model are given and the application of the model is outlined.

Chapter 5 describes the use of tables contained in Appendix 3 for the solution of economic life and replacement by model analysis. The development of the tables is discussed and their use is demonstrated by an example problem.

The importance of keeping adequate cost, time, and production records is stressed in Chap. 6. The development of a chart of accounts is described to enable the equipment owner to design his own system. A method of leveling costs and production records for the purpose of future projection is described and illustrated by example.

Equipment financing is covered in Chap. 7. Cash purchase, bank loans, leasing, and rental methods are discussed in simple, understandable terms. Illustrated examples show how the different methods operate and can be analyzed for comparison. Discounted cash-flow analysis is used for this purpose.

The subject of Chap. 8 is standardization. Specific examples of equipment standardization by make and model, engine family, and combinations thereof are given. Advantages and disadvantages of standardizing are discussed thoroughly.

Chapter 9 is about the management of the equipment inventory. A logical system for numbering the equipment is described as well as methods for keeping inventory records. The storage of equipment under varying conditions of use and weather is discussed followed by security of equipment and assignment of custody in order to maintain responsibility.

Maintenance management is explored in Chap. 10. The scope of maintenance is covered in three categories: preventive maintenance, repairs, and overhaul. A simple manual system for keeping records and scheduling preventive maintenance is outlined. Some of the advanced techniques of maintenance such as engine oil analysis, chassis dynamometer, and ignition analyzer are described. Lastly some of the problems of maintenance related to cannibalization of parts, abuse of equipment, and training of mechanics are mentioned.

The last chapter deals with equipment safety. This subject has achieved prominence through the passage of the Occupational Safety and Health Act of 1970 (OSHA). The legal aspects are discussed first, followed by the particular effects of the law in regard to construction equipment. Roll-over protective structures, fenders, brakes, and other aspects of the law related to new equipment and the retrofitting of old equipment are explained. Safety training and establishment of a safety program are suggested as ways of meeting the requirements of the law. The establishment of a viable policy in regard to equipment safety provides a fitting closure to the text.

It is anticipated that managers who have been out of school for 5 or 10 years, those who have not had sufficient exposure to engineering economics or the computer, and those who want to enhance their knowledge of the responsibilities of ownership will be best served by this book. Most of the nitty-gritty technical details of the technical procedures have been left out in order to have space to discuss the topics from the manager's point of view. Those who seek more of the details will find references which may be followed up for more specifics.

Of course, no book is complete or perfect. Any opinions expressed are those of the author, who also accepts the responsibility for any errors or omissions in the text. The assistance of his many friends in the equipment business, too numerous to mention by name, is gratefully acknowledged. Appreciation is also due to many of his colleagues on the faculty at Stanford University and other universities across the country, from whom so much has been learned.

JAMES DOUGLAS

1
Introduction to Equipment Policy

PURPOSE

The rapid advancement of equipment technology during the past several decades has imposed on the owner and equipment manager the additional burden of establishing sound policy governing the use of this equipment. Advancing technology has increased the size and cost of equipment; government tax policies have imposed increasingly sophisticated methods of tax computation; and economic factors such as inflation, obsolescence, and interest rates have complicated decision analysis of equipment problems. This book is intended for the use of all those having to know about the establishment of equipment policy. Contractors and students alike will find things explained in construction terminology which is familiar to them. Techniques are explained in simple terms which can be understood by the unsophisticated reader and are illustrated by numerical examples. Because of the large span of competence of potential readers, subjects will be developed from a simple base, increasing in complexity as skill and understanding are developed.

POLICY DEFINED

Equipment policy is the framework from which the operating principles of equipment management are derived. It is what you need to know in order to establish the doctrine for prudent management of the business. Without policy, decisions are made randomly and generally by intuition. Policy gives operations the discipline which permits the owner to optimize decisions in order to maximize his profits. The establishment of policy is the responsibility of top management, no less than of the president or chairman of the board. Although it may be developed by a committee or someone on the staff, it is so fundamental to the success of the enterprise that it must finally be determined by the person in executive control.

WHY ESTABLISH POLICY?

Policy must be established for the guidance of both management and the operating divisions of the firm. It is impossible for the company to move uniformly in the same direction and at the same speed without a clear and concise understanding of company policy. It would be like a quarterback calling signals to a football team when the team does not know the plays. Policy related to replacement, for example, must be understood by maintenance and operating personnel as well as by those in accounting. For the uniform guidance of the actions of all those who must make decisions regarding equipment, policy must be formulated, disseminated, and, above all, discussed.

The need for clear-cut construction equipment policy is amply indicated by an increase in the following factors:

1. Size and value of equipment fleets under the control of a single manager
2. Size of heavy construction contracts necessitating larger equipment investment on each job
3. Cost and complexity of individual machines resulting from technological progress
4. Availability of computer for cost keeping and investment analysis
5. Competition forcing contractors to adopt improved management methods in order to stay in business

The factors are a result of the modern economic environment, which seems to favor the large corporation. The larger companies are able to afford many highly trained specialists who are capable of applying the new technology to the construction business. In order to survive, smaller contractors must adapt these new technical resources to their operations. Computers, automation, and sophisticated analytical methods are here to stay.

The best way to establish policy is through the development of a set of instructions or a policy manual. In the small firm, the former would be enough, but in a large corporation, the manual is essential. Who writes this policy? Everyone must contribute. Perhaps a staff assistant will be designated to do the actual writing, but final approval must come from the top manager. Policy on standardization certainly should be developed with the help of the mechanics who maintain the machines and the operators who run them. The biggest single factor in the successful establishment of a viable policy is to obtain the cooperation and understanding (and agreement, if possible) of those people who will carry it out. A thorough discussion of the policy while it is being developed and especially the feeling of participation by the people involved in it will assist materially in its acceptance when it is promulgated by management.

ELEMENTS OF POLICY

A sound equipment policy must include all the elements which affect the ownership and operation of equipment. A proper consideration of engineering economy, finance, technology, ecology, and the law must be utilized in establishing policy. There is an ever-present conflict in making these decisions. A technically feasible machine may not be legal because of its size or weight. An economically superior machine may contribute so much to pollution that it is not acceptable. Each element of policy should be considered separately and then all should be merged to ensure compatibility. What are these elements?

There are many things which impinge on the ownership and operation of equipment. A study of policy should include a consideration of at the least the following items:

1. Economic life of various machines
2. Timing of machine replacement
3. Standardization of equipment
4. Degree of maintenance
5. Renting or leasing versus owning

6. Contract maintenance
7. Selection of equipment
8. Organization for administering equipment
9. Equipment analysis for cost and production
10. Cost records and operating records
11. Inventory management
12. Storage and security
13. Safety
14. Spare-parts support and stock levels
15. Air pollution, noise abatement, and creation of attractive nuisances
16. Public relations and sales psychology
17. Tax effects

While it is not possible to discuss all these elements in this book because of limitations in length, the principal ones will be covered. In particular, the subjects of economic life, replacement, standardization, and other elements with a flavor of engineering economics will be discussed at some length.

SOME EQUIPMENT STATISTICS

In order to grasp the magnitude of the equipment problem, it is necessary to develop some statistics on the national inventory of construction equipment. The best information on this subject comes from the archives of the U.S. Department of Commerce, generally obtainable from the "Statistical Abstract."

NATIONAL EQUIPMENT INVENTORY

Using the figures from Table 1161, Value of Shipments, Construction Machinery and Tractors in the "Statistical Abstract,"[1] and assuming a median age of 5 years, one can deduce that the national inventory of this equipment is about $40 billion, based on the original cost of the equipment. This does not include the passenger cars and highway trucks used by construction contractors. It is conceivable that the total value of equipment used in the construction contracting industry exceeds $60 billion.

It is regrettable that there is no national inventory of this equipment. It is strange indeed that all airplanes are registered by the Federal Aviation Administration in Oklahoma City, all

---

[1] U.S. Bureau of the Census, "Statistical Abstract of the United States 1970," p. 725.

highway vehicles registered by states in archives at the state capitals, all boats registered by the U.S. Coast Guard by districts, and yet there is no permanent record of our assets in construction equipment. This knowledge would be extremely valuable for emergency planning for catastrophes and even master planning to determine the feasibility of a massive dam, tunnel, or shelter program.

Unfortunate as it may be that we have no accurate knowledge of our equipment assets, it is clear that we have an enormous amount of it. Given this large inventory of equipment, it is easily discernible that sharper methods of figuring costs and maximizing profits and improved management will result in large differences in dollars.

CONSTRUCTION-EQUIPMENT OWNERSHIP

Periodically our national magazines reporting on equipment publish figures which give us an interesting insight on the amount of equipment contractors purchase and own. *Engineering News-Record*[1] stated that heavy and highway contractors in the top 400 in the United States had a median investment of $300,000 per $1 million of contracts in 1970. It is estimated that these contractors have invested more than $2.4 billion in machinery and trucks at current replacement value.

In the spring of 1971, *Construction Methods and Equipment*[2] surveyed more than 500 contractors with a total volume of business of about $19 billion and found that they owned more than $2.4 billion in equipment. Among the heavy-construction contractors surveyed who did $50 million and over per year in contracts, the median investment in equipment was $380,000 per $1 million. Purchases of new equipment for this group amounted to $52,000 per $1 million for 1970.

These statistics are from the past. Future projections of equipment values are even more impressive. The Federal Highway Administration[3] estimates that it will cost states, counties, and municipalities an average of $5.8 billion a year to maintain all the roads and streets in this country between 1973 and 1985. Equipment costs are estimated to be 27 percent of this total, or an average of $1.6 billion per year for the 12-year period.

These are impressive figures; combined with those of the

---

[1]*Eng. News-Rec.*, Apr. 8, 1971, p. 48.
[2]*Constr. Methods Equip.*, May 1971, pp. 15–16.
[3]*Eng. News-Rec.*, Dec. 4, 1969, p. 13.

construction industry, they indicate the magnitude of our present and future equipment investment. It is easy to see that savings of a small fraction of this investment will be in the millions of dollars. This, then, is the target for the establishment of sound equipment policy.

INFLUENCE OF TECHNOLOGY

No one can deny that technological progress during the past two decades has had a profound effect on construction equipment. It has affected production in dynamic ways and the life of the machine in more subtle ways. The allowance for normal obsolescence included in the depreciation allowance is a recognition of this fact. Policy is often confused because of the owner's inability to recognize obsolescence in his computations of economic life and replacement. For this we need to seek help from engineering economy, and this we shall do in a later chapter.

GROWTH IN SIZE OF MACHINES

The size of a machine usually refers to its gross weight or capacity (total volume). In earth-moving equipment, size is commonly measured either by struck capacity or tons of payload carried. The percentage rate of growth in size is the average percent increase in size compounded annually. When a new type of machine is invented, technology brings rapid change in the early generations of the machine. Improvements tend to reach a plateau in several decades, and changes then take place more slowly. This enables one to estimate the rate of growth by considering how long that type of machine has been on the market.

The automobile is a good example of the evolutionary character of a machine. Early cars had magneto ignition systems; few of these are found on cars today. The self-starter replaced the crank in the twenties, the automatic choke and hydraulic brakes came in the thirties, hydromatic transmissions came in the forties. Although there were many subtle improvements in metallurgy, electric systems, and the like during the sixties, there were no big jumps like those in the early days.

This is true of growth in the size of earth-moving machines. Figure 1-1 shows the evolution of Caterpillar scrapers in capacity. Over a 15-year period, they grew rapidly in size from 7 yd^3 stuck capacity to 40 yd^3. An upper plateau has been reached because of highway width, weight, and bridge limitations. Changes in highway laws may stimulate some small growth in

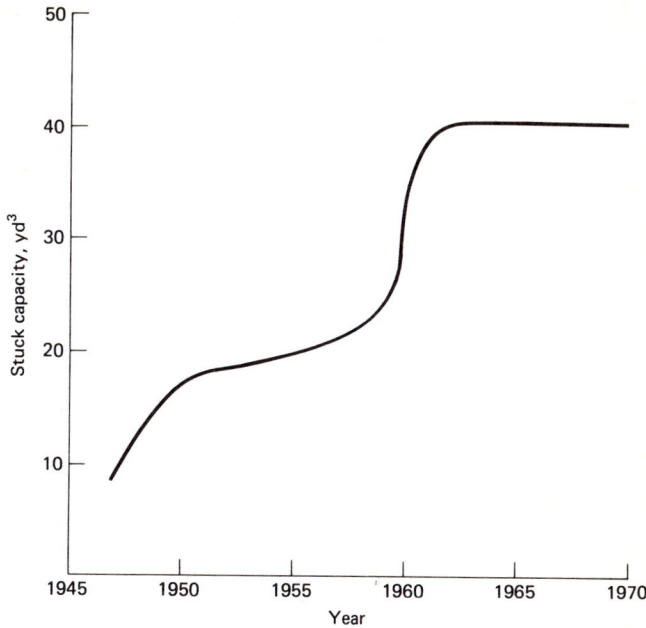

Fig. 1-1 Evolution of Caterpillar wheel tractor-scrapers in capacity.

the future, but width and height limitations defined by bridges, tunnels, and overpasses will not change materially in our lifetime. Perhaps the next increments in capacity will result from the design of automated earth-moving trains. Or perhaps a new earth-moving concept will replace the scraper in the same way the front-end loader is crowding the shovel off the market.

GROWTH IN PERFORMANCE OF MACHINES

During the 15-year growth of scrapers in stuck capacity, an even more remarkable growth was taking place in performance. Both horsepower and speed increased faster than size. Figure 1-2 shows the evolution of Caterpillar scrapers in horsepower per ton of gross weight, a good indication of increased speed and acceleration which result in better performance. Figure 1-3 shows the increase in maximum speed of Caterpillar wheel tractors over the same period. Both figures show that performance has more than doubled in the span of 15 years. Horsepower per ton increased from 3.3 to about 8, and top speed jumped from 20 mi/hour to over 40 mi/hour for the four-wheel tractors.

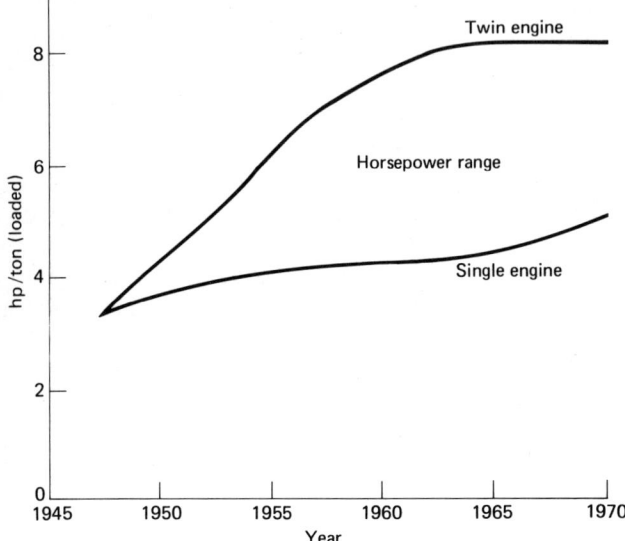

Fig. 1-2 Evolution of Caterpillar wheel tractor-scrapers in horsepower per ton.

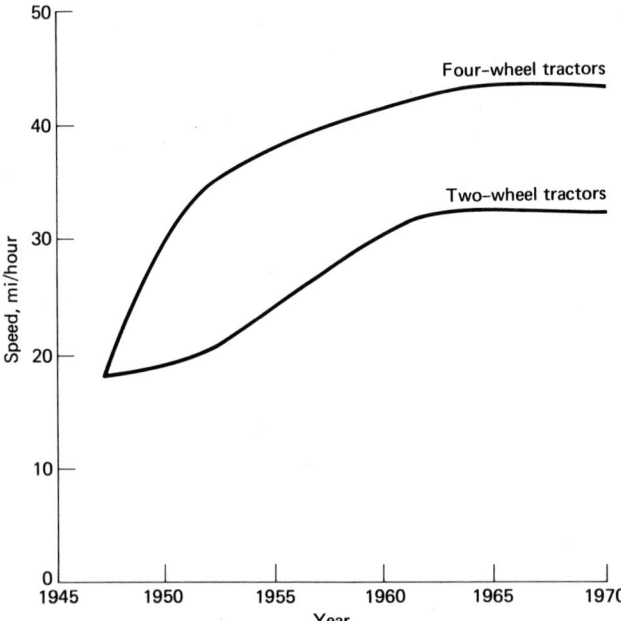

Fig. 1-3 Evolution of Caterpillar wheel tractor-scrapers in maximum speed.

INTRODUCTION TO EQUIPMENT POLICY

RAPID TECHNOLOGICAL CHANGE MAKES ANALYSIS DIFFICULT

Before the Industrial Revolution, changes in equipment and materials took place very slowly. It is interesting to contemplate the speed at which man has traveled over the historical past. Figure 1-4 clearly shows the changes wrought by the advent of technology. Until about 4000 B.C. man's speed was limited by how fast he could run. A 100-yd dash run in 10 seconds is a speed of 20 mi/hour. A 1-mi run in 4 min is 15 mi/hour. But since it is impossible to sustain these speeds over any substantial distance, man might be credited with a speed of 5 to 10 mi/hour, depending on the distance and conditions.

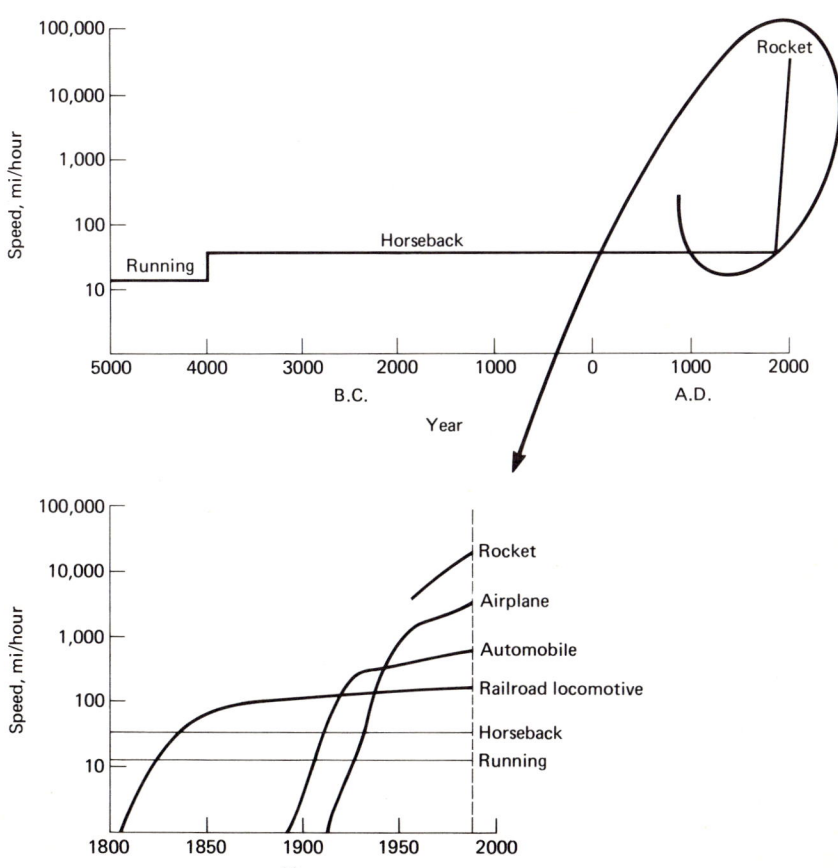

Fig. 1-4 Evolution of speed of travel.

Records of the past indicate that man succeeded in getting the horse to carry him at speeds faster than he could run about 6,000 years ago. A good racehorse can run a mile at about 30 mi/hour. For greater distances, the average speed of the horse will be between 10 and 20 mi/hour, say 15 mi/hour. For 6,000 years, man was constrained by the "hay barrier," the speed at which a horse could travel.

Although the Industrial Revolution was a child of the eighteenth century, real progress came only in the nineteenth century. The first practical steam locomotive was invented in 1803, signaling a new era of locomotion. By 1829, it had achieved a speed of 24 mi/hour. In 1837, the steam locomotive North Star rushed down the rails at 60 mi/hour, the first to do so. Today, trains have achieved speeds of 200 mi/hour.

The next step forward came with the invention of the automobile about 1900. From an early crawl at 5 mi/hour, cars have progressed to speeds of 100 mi/hour or more for the ordinary car on a highway. Racing cars commonly achieve 200 mi/hour on the track, and jet-engine autos have made over 600 mi/hour on the salt flats in Utah.

The real technological breakthrough came when man took to the air. It was 100 years after the first practical locomotive that the Wright Brothers made the first powered flight at Kitty Hawk, N.C. (1903). C. P. Rodgers flew from New York to Pasadena in 59 days in 1911. By 1928, Arthur Goebel was able to fly from Los Angeles to New York in 18 hours, averaging 139 mi/hour; but by the end of World War II W. H. Council did it in 4 hours and 13 min at an average speed of 560 mi/hour.

Rocketry became practical with the advent of the German V-2 rockets fired at England in World War II. They traveled at something less than 2,000 mi/hour. The Russians launched Sputnik in 1957 to crack the space barrier at approximately 15,000 mi/hour. Today, speeds of over 25,000 mi/hour are relatively common in space flight. And all this has been accomplished in less than 200 years of the new technology.

Before the Industrial Revolution, the life of a tool or piece of equipment was determined by when it wore out. There was relatively little obsolescence to influence its life span. In the modern economic environment created by the Industrial Revolution, new criteria must be used to make life and replacement decisions. Inflation, taxes, and interest rates, as well as obsolescence, all influence these decisions. Technology has thus forced equipment

INTRODUCTION TO EQUIPMENT POLICY 11

owners to adopt more sophisticated methods for implementing equipment policy.

SOME SURVEY RESULTS

With the inception of the Gallup Poll several decades ago, surveys became a popular method of determining how people feel about issues. It is natural that we should turn to surveys as an aid in finding out what is troubling the contracting industry in the construction-equipment policy field. Several polls have been conducted by the American Society of Civil Engineers' (ASCE) Construction Equipment Committee and the Construction Institute at Stanford University.

ASCE SURVEYS OF CONTRACTORS AND MANUFACTURERS

In April 1960 the Construction Equipment Committee made a survey of construction contractors to find out what their problems were in regard to construction equipment. Eighty-six replies were received, representing about a 20 percent return on the questionnaires mailed out. Respondents included 14 presidents, 26 vice-presidents, 20 superintendents, 8 managers, and 5 engineers, indicating that the survey was rewarded with high-level attention.

The major equipment problems confronting the industry at that time, listed in order of priority, were:

1. Programming preventive maintenance
2. Training and motivating operating and maintenance personnel
3. Determining optimum equipment life
4. Influencing standardization and improvement
5. Determining maintenance costs
6. Controlling spare-parts inventories

It is inconceivable that these problems have disappeared in the last decade. There has been entirely too little research on them to expect any real improvements. Communication is difficult, too, for there is so very much of it in the modern business world. What has been written about these problems has become lost in a morass of trivia on how to solve small problems.

In May 1961 a follow-up survey of manufacturers was conducted by the ASCE to find out how they could help solve the problems enumerated in the first survey. Out of 200 equipment manufacturers contacted, only 7 percent responded to the questionnaires. The replies received were nonspecific, and some were frank in admitting they had nothing to offer. There was however a general willingness to cooperate among those who responded.

It is interesting to note that many, if not most, of these problems are people-oriented. They relate directly to the attitude, motivation, training, and organization of the personnel who relate to the equipment. This sort of problem cannot be solved without working on the people end of the equation as well as the equipment.

Most major manufacturers noted that they operate training schools and assist in keeping maintenance mechanics up-to-date through correspondence courses and schools at the factory or training center. One comment was that design engineers seem to have developed complex components and devices more rapidly than people have been trained to use and maintain them.

The last major classification of comments included in this report was the manufacturers' suggestions on action which could be taken by the users of equipment. Here the manufacturers fired their biggest salvos! Several pointed out that proper selection of equipment for a given job, standardization, and compliance with existing recommendations on lubrication and maintenance were, are, and properly should be a user responsibility. One suggested that what contractors needed was a program for the utilization of literature already available from the manufacturer. Typical comments in this area were:

> Our experience indicates that in many cases these [operating and maintenance] manuals, which as a rule are very complete, are neither read nor preserved by the operators. Even such simple instructions that appear on instruction plates secured to the machines themselves are frequently disregarded.
>
> In some instances operators buy their vehicles on the basis of low dollar which can mean undersize components or even components of approximately correct size but not of the type which will give the required reliability. Low dollar equipment is produced from low dollar components which frequently do not give the intended service life.

Sound familiar? People problems. Another reason why equipment policy needs the unified support of all levels of management.

THE STANFORD SURVEY

During the months of August to October 1962, a survey of construction equipment ownership policy was conducted under the auspices of Stanford's Construction Institute. Firms participating were, for the most part, members of the Construction Institute. A total of 36 firms responded with completed questionnaires. These were followed up, in most cases, with personal interviews of selected executives. In several firms, two or three branch or district offices of the same company were contacted.

The one common denominator in selecting firms for interview was an interest in construction-equipment ownership policy. The survey spanned the construction industry, from dealer to user, from small to large. Responses are indicative of what the industry is doing and what practices are followed. These practices may not always be the optimum policy. Some conclusions reached in this survey were:

1. Each firm must determine the best policy for it under a given set of circumstances. No golden rule seems to fit all firms at all times.
2. The useful service lives of various types of equipment are subject to many variables. An infinite number of combinations of utilization, maintenance standards, job conditions, and operating policies provide large differences in the lives used by different firms. In general, however, the survey indicates that lives of *more than 5 years* should be examined carefully for justification.
3. There is a common tendency to rely on judgment or intuition where an engineering-economy approach is valid. Reduction of intangibles to a probable dollar value would be of great assistance in the analysis of complex problems.
4. Most cost figures kept by contractors are incomplete and not entirely reliable. Few contractors collect annual hours of operation on their equipment with any accuracy.
5. There is no common understanding of terms used in the industry. There should be standard definitions for such terms as operating hour, utilization, availability, etc.
6. There is no clear understanding of obsolescence. At least half the respondents believe the annual rate of obsolescence is 20 percent, and another quarter think it is more.

These conclusions are the result of a careful study of equipment policy by experts in the field. There is a hint of uncertainty in all these findings. Are we really sure we understand the basic tenets of policy related to construction equipment? Since 70 percent of those questioned stated that they used judgment or intuition in making decisions on equipment replacement in lieu of economy studies, a basic knowledge of these techniques in the industry is doubtful.

ELEMENTS OF EQUIPMENT POLICY

There are a large number of things which can be included in equipment policy, but the most important of these must surely be:

1. Inventory and asset account keeping (including classification)
2. Standardization within the company
3. Cost keeping and cost records
4. Time and production records
5. Renting and leasing
6. Replacement
7. Economic life
8. Maintenance practices and records
9. Tax effects on ownership

A clear enunciation of policy related to the above elements is necessary for the healthy survival of the construction contractor. How this policy is disseminated and what depth it reaches will depend very much on the personalities of the people in the firm. It will be necessary in all but the smallest enterprises to develop this policy in writing and to communicate it orally below certain levels. Whatever form the communication takes, it must be done positively in order to reach all levels, including the lowest.

2
Equipment Economics

DEFINITION

Equipment economics is the science that deals with the management and utilization of equipment. It must be firmly based on the principles of engineering economy clearly enunciated in Grant and Ireson's *"Principles of Engineering Economy"*[1] or an equivalent text. A good grasp of the principles is fundamental to sound economic analysis of equipment. A brief review of these basic ideas will bring the reader up current and show how they apply to equipment problems.

THE IMPORTANT PRINCIPLES

The important principles to be considered and utilized in the analysis of equipment problems are:

[1] E. L. Grant and W. G. Ireson, "Principles of Engineering Economy," 5th ed., Ronald, New York, 1970.

1. Separable decisions should be made separately.
2. It is the difference between alternatives that matters.
3. Money has a time value.
4. Items should be given a money value to make them commensurable.
5. Sunk costs are immaterial except as they affect future costs.

A proper consideration of these principles in relation to any equipment problem will serve to clarify thinking, simplify the ultimate resolution of a problem, and guide the analyst into arriving at the best solution. Let us now consider the application of these principles in more detail.

SEPARABLE DECISIONS

It is very common in considering the complexities of a problem to confuse and combine decisions which can and should be separated. Separation of a problem into its component parts usually will simplify the problem and make it easier to solve. In doing this, care must be taken to recombine the parts to get an overview of the solution and its effect on the entire system. If one does not take this final step, it is possible to suboptimize. Suboptimization may result in the best solution for one part of a problem at the expense of other parts.

Generally speaking, the final answer must meet the test of being technically correct, economically sound, financially feasible, and politically acceptable. And these are decisions which must be made separately. Sometimes the best alternative from the economic viewpoint may not be politically expedient and must be rejected in favor of an alternative less sound economically but more acceptable in a political sense. Emphasis on environmental results of decisions will make this facet of decision making more important in the future than it has been in the past.

When the available alternatives have been examined and tested to meet the criteria above, the final decision will probably entail a value judgment of the factors involved. While the computer and other machines may be of considerable assistance in analyzing a problem, the final decision will always have to be made by one individual who can evaluate the irreducibles in the problem. Giving weight to each part of a problem involves a consideration of the interrelationship between the various parts, the systems approach. In finality, wherever decisions can be made separately, this should be done, with a final recombination to test the solution.

DIFFERENCES BETWEEN ALTERNATIVES

After a problem has been defined, several alternative solutions are generally found to be available. Invariably it is the differences between these alternatives which must be sought out and quantified in order to decide which is the best solution. As a rule, you must measure the differences between alternatives, for that is what really counts. Remember, too, that these differences may change in time, so that it will be the differences as they appear to be at the time action finally is taken which will be pertinent. Where differences are likely to change between the time of analysis and the time of final action, these differences must be forecast and included in the study.

Example A contractor requires the services of a portable air compressor on a job expected to last 1 year. He is able to buy a new diesel-engine-driven rotary 365 ft^3/min compressor for $16,830. He estimates that in the future market he can sell the machine 1 year from now at a price of $12,500. The estimated cost to operate this machine will be $2.65/hour, and he anticipates 1,500 hours of use in the next year. The dealer offers to rent him a new machine of the same type for $590/month or $200/week. What should the contractor do? Use 0 percent interest rate.

Solution The capital cost of the new machine will be

Cost new	$16,830
Salvage value	12,500
Capital cost	$ 4,330

The cost of operating this machine for the next year will be 1,500 × $2.65 = $3,975. The cost of renting the same machine for 1 year will be 12 × $590 = $7,080. Since the operating cost of the machine must be paid by the user, the cost of operating the rental machine will be the same as for the one purchased. Hence the difference in cost to be considered will be:

Capital cost to own	$4,330
Rental payments	7,080
Difference in cost	$2,750

The difference of $2,750 is in favor of owning the machine rather than renting it. Because the operating cost will be the same either way, its cost may be eliminated from the comparative figures. It is well to consider the additional risk which may be involved in the ownership of the machine. The new cost of the machine and the rental price are pretty firm, but the future salvage value of the machine is a cost 1 year from now. If there is a likelihood that the market value of this machine a year hence might be less than $12,500, this should be considered as an additional risk of ownership. Given the present figures, however, ownership is favored over rental.

In an economy study, there will almost always be several alternatives. As a first consideration in equipment-economy studies, there will be the alternatives of *with* or *without* a given machine. As a least consideration, all alternatives should be considered and evaluated unless obviously infeasible. There will be times when the least likely contender among them will turn out to be one of the best. In comparing the alternative costs of various machines, it is not the first cost of a machine that really counts but the total cost of the machine over a lifetime of useful service.

The replacement decision is basically one of considering several alternatives: there is at least a difference in the cost of replacing now versus replacement at a later date (1 year is a common future date for analysis). Then there may be additional alternatives involving several competitive replacement machines. Insofar as possible, all replacements should be considered, in size and numbers as well as make and model.

It is the careful consideration of all alternatives that bears the best fruit in equipment analysis. Whatever the alternatives are, there will certainly be differences in the timing of costs. For that reason, the time value of money should be reckoned.

TIME VALUE OF MONEY

Money has a time value. By this is meant that the possession and use of a certain amount of money generate a charge (cost) against that money. It has been well said that a firm is hemorrhaging money from the date of its beginning. This is symbolic of the fact that equity capital employed in the establishment of an enterprise must generate a return or be profitably employed elsewhere. One point of view is that the return generated should be at least equal to the cost of bank borrowing. In view of the risk involved in a private venture, the return would probably have to be greater in order to be attractive.

When a bank lends money, it makes a time charge which is called interest. Interest charges are made at the end of a time period (usually a year). These interest charges may be considered as a rental rate for the use of the money. When these interest charges are made at the end of the time period and added to the principal, the total amount will bear interest at the end of the next time period and the procedure is known as compounding. Compound interest is the interest rate charged on the total amount of principal and interest for the previous time period. If the amount of money borrowed is called the principal P and the compound interest rate is i, the amount due A at the end of n years will be

$$A = P(1 + i)^n$$

The term $(1 + i)^n$ is called the compound amount factor and can be found in interest tables by use of i and n. The amount A might be called the future worth of the principal P at the end of year n and at an interest rate of i. Compounding may not be at yearly intervals but may be semiannually, quarterly, or monthly. A common practice with department-store bills today is to compound monthly at $1\frac{1}{2}$ percent interest. This amounts to a nominal annual interest rate of 18 percent (12 × 1.5) and an effective rate of 19.6 percent $[(1.015)^{12} - 1]$.

Sometimes add-on interest is used in loan computations. This is quite common in automobile and equipment loans and leasing arrangements. The usual procedure in computing add-on interest charges is to use the full amount borrowed over the full term of the loan. When principal and interest are paid back in monthly installments over a period of several years, the equivalent compound interest rate will be approximately double the simple interest rate used in the computation. One must be very careful in evaluating interest rates quoted in this fashion.

Example A contractor borrows $3,600 at 8 percent add-on interest to buy a small engine-driven generator set. He pays back the principal and interest in 36 equal payments (3 years). What is the nominal rate of interest compounded annually?

Solution Monthly interest payments amount to ($3,600 × 0.08)/12 = $24. His monthly payments will be $100 + $24 = $124. By interpolating in the interest tables and use of present-worth factors for a uniform series of payments, the monthly rate of compounding can be found to be 1.21 percent and the nominal annual rate of compound interest to be 12 times that, or about 14.5 percent, and an effective rate of about 15.1 percent. This compares unfavorably with the 8 percent add-on interest rate in the computations.

The rate of interest to be used in compounding is a matter of some argument. As evidenced by past reports, many institutions do not use any rate of compounding in their economy studies. To ignore interest charges (i.e., use 0 percent rate) is to overlook the realities of life, even in governmental institutions. While the rate used by a branch of the government may be low, say 4 percent, it should never be lower than the present rate of long-term borrowing, such as the interest rate on bonds issued by the agency. Some consider a minimum rate to be the present bank-loan rate, say 10 percent. Others feel that it should approach the minimum attractive rate of return on investment in the enterprise. This may be 20 percent or more. The higher the rate, the greater will be the emphasis placed on present and near dollars, the less on dollars of the distant future. Whatever nominal rate is used, it should be increased to cover the degeneration of capital assets through inflation. A good minimum value of interest rate to use in equipment studies is 10 percent. If the inflation rate is added to that to obtain an operating minimum, a rate of 15 or 20 percent can be easily justified.

The kind of compounding described previously is known as incremental compounding because it is accomplished at equal increments in time. It is sometimes useful and convenient to resort to continuous compounding. For example, in business it is common to receive and disburse funds daily. This can be treated as a problem in continuous compounding. It can be demonstrated mathematically that the continuous compound amount factor is equal to e^{rt}, where $e = 2.718$, r is force of interest, and t is time in years. The variable r is called the force of interest in continuous compounding, and for values of less than 10 percent it is very nearly equal to the incremental rate of interest. Problems in continuous compounding can be solved easily by use of a log-log slide rule. Table 2-1 is a list of effective interest rates and their equivalent forces of interest.

While the technique of evaluating future dollars is useful in understanding the theory of compounding, a more useful device in making economy studies is the computation of present value by a reversal of the computations. The equation for determining the future value of a certain amount of money is

$$A = P(1 + i)^n$$

and the same equation can be restated to determine the present value of a future sum. In this case, the equation is

$$P = \frac{A}{(1 + i)^n}$$

where P = present value (or present worth)
A = future value in n years at annual rate of interest i

EQUIPMENT ECONOMICS

Table 2-1 Effective interest rates and equivalent forces of interest

Effective interest rate,%	Force of interest,%
1	0.995
2	1.980
3	2.956
4	3.922
5	4.879
6	5.827
7	6.766
8	7.696
9	8.618
10	9.531
15	13.976
20	18.232
30	26.236
40	33.647
50	40.547

The factor $1/(1+i)^n$ is the single-payment present-worth factor, also called the discount factor, which is used to determine the present worth of future dollars. This factor is the reciprocal of the compound amount factor and can be found in any good set of interest tables by knowing the values of i and n. It also can be computed quite easily on a log-log slide rule.

It has been mentioned that the effect of interest in discounting is to emphasize the value of present and near dollars while diminishing the effect of dollars in the future. By discounting at a 10 percent rate of interest it can readily be determined that the present worth of a dollar 20 years from now is about 15 cents and that of a dollar 40 years from now is only 2 cents. If you deposited a dollar at the end of each year from now into the infinite future, the present worth of the infinite series of deposits would be $10 and about 75 percent would be acquired in the first 15 years.

Since discount factors are very useful in analysis by discounted cash flow, a simple problem will be solved to demonstrate this method.

Example A contractor purchases a 1¼-yd³ crawler type front-end loader for $20,000 cash. He estimates this machine will earn revenues of about

Table 2-2 Computations for present worth

End of year (1)	Cash out (2)	Cash in (3)	Difference (4) = (3) − (2)	Discount factor (5) = $1/1.1^n$	Present worth (6) = (4) × (5)
0	20,000		− 20,000	1.000	− 20,000
1	20,000	30,000	+ 10,000	0.909	+ 9,090
2	21,000	30,000	+ 9,000	0.826	+ 7,434
3	22,000	30,000	+ 8,000	0.751	+ 6,008
4	23,000	30,000	+ 7,000	0.683	+ 4,781
5	24,000	30,000	+ 6,000	0.621	+ 3,726
			+ 20,000		+ 11,039

$30,000/year for the next 5 years. Maintenance and operating costs (including operating labor) are estimated to be $20,000 the first year, increasing $1,000/year after that. Neglecting the many associated costs except for maintenance and operation, what is the present worth of this investment at 10 percent interest rate? What is the return on the investment?

Solution The best way to solve this problem is to set up a table, fill in the table with the known values, and solve for the unknown values by slide rule (or desk-top computer). The procedure for solving this problem is known as the discounted-cash-flow method because cash flows are discounted back to the present (time 0). Table 2-2 shows the solution for the first part of the problem, which is to determine the present worth of the investment at a 10 percent interest rate. From the result of the computations in this table, it is seen that the present worth of the investment at 10 percent interest is $11,039. Note that the total difference between cash in and cash out, without interest consideration, that is, 0 percent interest, is $20,000. The application of interest reduces the present value of the future dollars.

The answer to the second question of the problem can be found by increasing the interest rate until the present worth of the cash flow is zero. This will be the return on the investment. Solution can best be accomplished by trial and error with a final interpolation when you are close to the answer. Table 2-3 shows the computations for return. It will soon be

Table 2-3 Computation for return on investment

End of year (1)	Difference (2)	30% discount factors (3)	Present worth at 30% (4) = (2) × (3)	35% discount factors (5)	Present worth at 35% (6) = (2) × (5)
0	− 20,000	1.000	− 20,000	1.000	− 20,000
1	+ 10,000	0.769	+ 7,690	0.741	+ 7,410
2	+ 9,000	0.592	+ 5,328	0.549	+ 4,941
3	+ 8,000	0.455	+ 3,640	0.406	+ 3,248
4	+ 7,000	0.350	+ 2,450	0.301	+ 2,107
5	+ 6,000	0.269	+ 1,614	0.223	+ 1,338
			+ 722		− 956

found by working up in 10 percent increments where the solution lies. A 5 percent bracket of the answer is good enough for interpolation. In this case, interpolation yields

$$\frac{722}{1,678} \times 5\% = 2.15\% \text{ say } 2\%$$

The return on the investment is 32 percent.

Since not all costs were considered in working up this return, it is known as a relative rate of return. Had all costs been included, it would have been an absolute rate of return. Great care must be used in analysis by relative rate of return, for the result will be responsive to what costs have been included (or conversely, what costs have been left out).

In reckoning costs by use of compound interest certain conventions are generally followed to simplify the computations. Usually these assumptions do not affect the outcome materially. Where it is suspected that they do, more accurate computations should be made using factual data. One of the most common conventions is the assumption that all costs or charges occur at the end of a year. Interest tables are made out based on this convention. While this device may not be entirely accurate where daily receipts and monthly payments are being considered, results are generally in line with the accuracy of other forecast conditions. As mentioned previously, the difference between the force of interest in continuous compounding and the rate of interest in incremental compounding is small for values of less than 10 percent.

Another convention often used in depreciation accounting is the half-year convention, which assumes that acquisition and disposal of equipment occur at the end of a half year. If the fiscal year adopted by a firm is the calendar year, all equipment bought and sold during that year would get ½-year depreciation expense based on the assumption that the change took place on July 1.

Treasury Decision 7128[1] established a modified half-year convention for depreciation accounting where it can be assumed that a machine acquired at any time during the first 6 months of a year will be allowed a full year's depreciation expense and one acquired at any time during the last 6 months will be allowed ½-year depreciation expense.

The time value of money is a fact which must not be ignored. Interest is rent charged for the use of money and is generally

---

[1] U.S. Treasury Dept., *Treas. Decis.* 7128, 1971.

levied at a rate dependent on the supply and demand for risk capital and other economic factors. The rate of interest used is variable and depends on the use of the funds, the availability of money, inflation and other economic factors, and the nature of the enterprise. Interest rates should generally lie between 4 and 20 percent. Anything outside this range should be used only with a full consideration of the factors involved. As a general rule, a 10 percent rate of interest will be satisfactory in the study of equipment economics except for governmental agencies, where 6 percent might be more acceptable.

GIVING ITEMS A MONEY VALUE

One of the important things about economic analyses is to measure the differences between alternatives. These differences are very hard to measure unless they can be quantified. In analyzing equipment, items should be given a money value whenever possible in order to make them commensurable. Items which cannot be quantified are called irreducibles and should be set aside for consideration when economic calculations are completed.

A good example of this type of item which is not ordinarily quantified is the penalty charge for equipment breakdown. Consider the case of a transit mixer delivering concrete to a construction job. Halfway to the job, the fuel pump fails, and the truck stops beside the road. A frantic radio call to the dispatcher will get a mechanic in a pickup truck to the transit mixer in half an hour. It will probably take 2 hours more to locate the trouble, get a new pump, and replace the old one on the engine. Meanwhile, it is a hot day and the concrete has set in the mixing drum. The concrete crew has been standing idle on the job for several hours awaiting the delivery of concrete. The contractor already has thoughts about backcharging the labor of the concrete crew to the supplier. No one is happy. Some of the direct costs of the repair will undoubtedly be charged against the truck that broke down. The new fuel pump and the labor to replace it will usually be the extent of these charges, along with some overhead. But how about the mechanic's time in transit and the cost for use of his pickup for a couple of trips to the site of the breakdown? How about the cost to clean the concrete out of the mixer barrel? The cost of the contractor's concrete crew? The disruption of an orderly work schedule for both dispatcher and mechanic? And the loss of goodwill? Most of these costs can be searched out and quantified. They should be collected in a

EQUIPMENT ECONOMICS

special account for breakdown or penalty costs and used in any economic study to determine life and replacement decisions. If the analyst does not feel competent to evaluate the loss in goodwill, perhaps the owner can help.

Any items which cannot be evaluated in terms of cost, i.e., irreducibles, should be set aside for consideration after cost computations are made. In the case of the transit mixer above, the loss of goodwill might be treated as such. The number of breakdowns could be collected and used at the time of decision. If two different makes of trucks compare favorably in costs, a consideration of the history of breakdowns could indicate clearly which is preferred as a replacement.

When it is difficult to quantify costs with precision, a sensitivity analysis is sometimes helpful. In checking the sensitivity of an economic model to a variable which is difficult to quantify, it is often possible to select the upper and lower bounds with some confidence. The model is solved with the values, and perhaps some in between these limits, to see what happens when these values change. Figure 2-1 is an example of this sort of

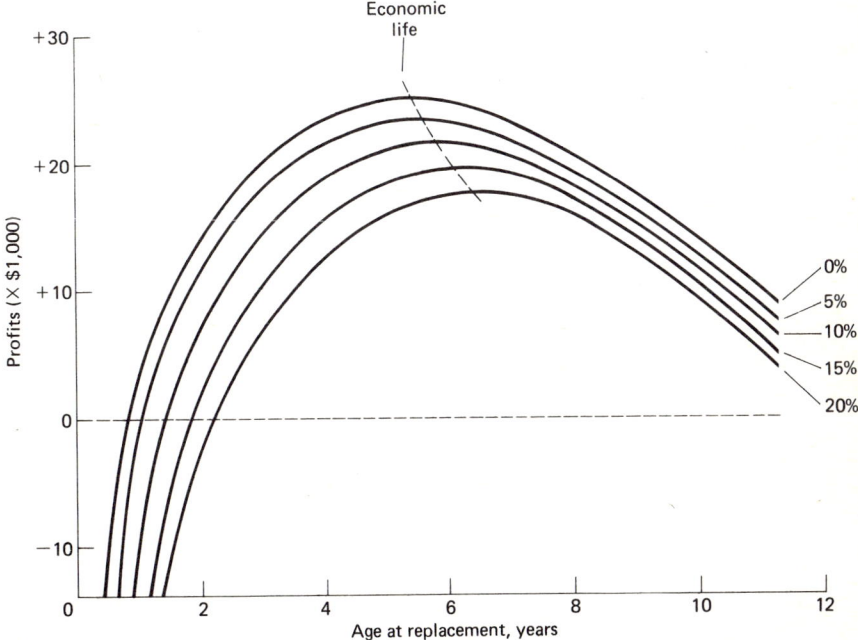

Fig. 2-1 Economic life and profits with inflation rates from 0 to 20 percent.

analysis. Suppose you are curious to learn the effect of inflation on profits and economic life of a crawler tractor. You feel that the annual inflation rate will be at least 5 percent but probably no more than 20 percent. With 0 and 20 percent as limiting values and 5 percent increments to get inside values, a set of curves is plotted showing profits versus economic life for the several rates. It will be observed that profits decrease and economic life increases as the inflation rate increases. This is a typical sensitivity analysis.

In handling certain irreducibles such as breakdown costs which may be difficult to quantify, upper and lower bounds can be estimated and the sensitivity of the model checked against these values. It may be that economic life is not very sensitive to this variable. If so, the differences between alternatives will generally be insensitive to it also. In this case, a median value of the variable is probably good enough. If it should turn out that the model is highly sensitive to the "irreducible" variable, more time and effort should be expended to evaluate it. If this does not yield the expected results, a subjective judgment should be made at the end of the analysis as to the impact and likelihood of a change in value on the decision.

SUNK COSTS AS A FACTOR

Sunk costs are past costs. Unlike irreducibles, sunk costs are known since they have already occurred. Since they represent money spent that cannot be recovered, they should be disregarded except where they affect future costs. Money spent for a major engine overhaul, for example, cannot be fully recovered, but it certainly affects the future maintenance and operating costs of the machine.

Since economic analyses involve future costs and a comparison of alternatives, sunk costs will have the same effect on all alternatives. Since it is the differences between these alternatives which really count, the sunk costs are generally irrelevant to the study. It is this consideration of future costs which makes the equipment analysis difficult. In many ways it is like estimating the cost of construction work. Historical costs are important only as they serve as a guide to the prediction of future costs.

> **Example** Two years ago, Smith purchased a new portable air compressor for drilling rock. The cost of the compressor was $8,000. He made a down payment of $2,000 and has been paying $1,500 a year plus 8 percent interest on the unpaid balance. He still owes $3,000 on the machine, which, when paid off, will give him a clear title to it. If he defaults, however, the loan will

be recalled and the dealer will repossess the compressor. Smith has just signed a contract for a new job where a compressor of this size is required. Since purchase of the machine 2 years ago, the economy has slumped and the prices of new and used machines similar to it have dropped markedly. Jones, a friendly contractor in dire financial straits, offers to sell Smith an equally good compressor of the same age for only $2,500. What should Smith do? Should he acquire title to the machine on which he has already paid $5,000 plus interest charges, or should he return it to the dealer and pick up Jones's machine for $2,500?

Solution Since the replacement machine offered by Jones is equal to the machine Smith now owns, the future maintenance and operating costs of the two machines are estimated to be the same. The capital costs will differ, however, because Smith still owes $3,000 (2 years more at $1,500) plus interest on his present machine. The $5,000 Smith has already invested in his present machine is irrelevant in the choice which must be made. Obviously, it will be to Smith's economic advantage to return his present machine on which he owes more than $3,000 and pick up Jones's machine for $2,500. Smith probably has a strong feeling that since he already has invested $5,000 in his machine, he cannot afford to give it up in favor of another just like it. The decision must be based on sound economic justification and not on sentiment.

This problem has illustrated the general irrelevance of sunk costs and the importance of future costs in economic analysis. Let us now turn to some of the other factors which affect equipment economics.

3
Some Economic Influences

There are a number of economic factors which should be considered in equipment analysis. Taxes, inflation, and equipment obsolescence all have an effect on decisions that relate to the economics of the equipment.

One of the causes of the American Revolution was taxation without representation. The Bible is full of references to tax collectors. It must be presumed that civilized societies cannot avoid taxation of one form or another. Taxes will continue to be imposed in the future, perhaps even more burdensome than they are today. The taxes on personal and corporate income have a great influence on the economics of equipment and should be considered in its evaluation. Personal property taxes, licenses, and fees should also be accounted for, but their effects will be minor compared to the effect of income taxes. The three mediums through which income taxes will be affected are depreciation, investment credit, and gain on sale. Let us examine these items.

DEPRECIATION AS A TAX INCENTIVE

Depreciation is the decline in capital value of an asset because of exhaustion, wear and tear, and obsolescence. The law allows a reasonable allowance for depreciation to be taken as an expense of doing business. This expense is deductible from gross income for the determination of taxable income and consequently affects the amount of tax paid. For a more thorough treatment of the subject than will be found here, the reader is referred to Grant and Norton's excellent book[1] and the numerous publications from the Government Printing Office on the subject.

There are four principal elements in depreciation accounting: initial cost, salvage value, useful service life, and depreciation method. The most common methods of depreciation used currently for construction equipment are straight line, declining balance, and sum of years' digits. These are illustrated in Fig. 3-1, where the initial cost is $10,000 and the salvage value at the

[1] E. L. Grant and P. T. Norton, Jr., "Depreciation," Ronald, New York, 1955.

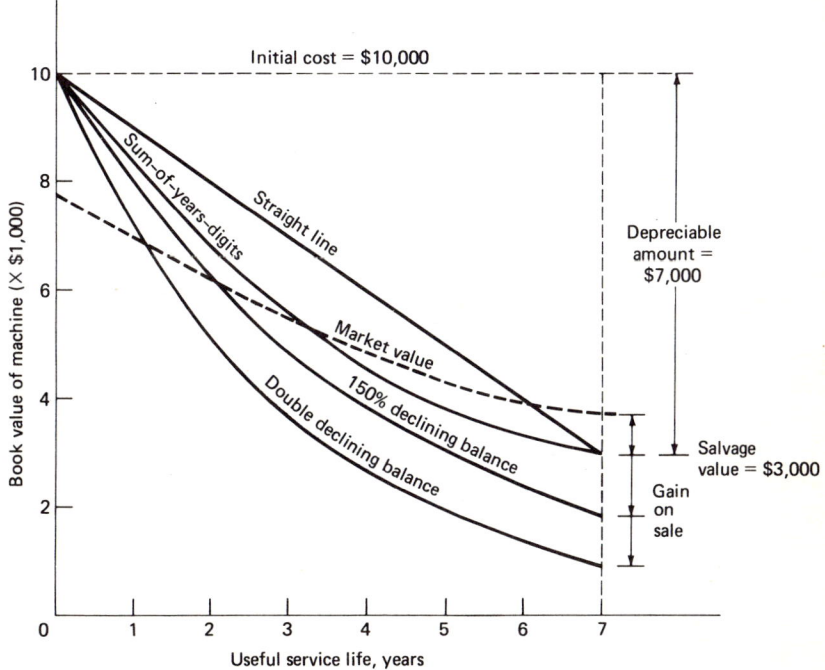

Fig. 3-1 Depreciation methods.

end of a useful service life of 7 years is $3,000. The difference between the two is $7,000, often referred to as the depreciable amount. The method of write-off determines the shape of the capital value (or book value) curves labeled straight line (SL), sum of years' digits (SOYD), and double declining balance (DDB). If the sale price of the used machine is not exactly equal to the book value (and it almost never is), a loss or gain on sale results. If all the depreciation is written off in the year purchased (usually to a salvage value of zero), the item is said to be *expensed*. If it is written off over a useful service life of several years, it is said to be *capitalized*.

Straight-line depreciation is the result of equal annual amounts being written off as depreciation expense. In Fig. 3-1, the depreciable amount of $7,000 divided by a life of 7 years results in an annual depreciation expense of $1,000. If these annual expenses are discounted back to the present, as shown in Table 3-1, the present worth of the straight-line depreciation is found to be $4,868 at 10 percent interest. Straight-line depreciation is one of the oldest depreciation methods. Its principles are easy to understand and apply. It is preferred by many taxpayers for those reasons.

Sum-of-years'-digits and double-declining-balance depreciation are known as accelerated methods and were established by the Revenue Code of 1954. They offer the economic advantage of a faster write-off than straight line but can be used only on new equipment which has a useful service life of more than 3 years.

Sum-of-years' digits depreciation is calculated by first finding the sum of all digits in the estimated life of the machine. Using the data of Fig. 3-1, the sum of digits for a life of 7 years is

Table 3-1 Straight-line depreciation

End of year (1)	Depreciation expense (2)	Present-worth factor at 10% (3)	Present worth (4)
0		1.000	
1	$1,000	0.909	$ 909
2	1,000	0.826	826
3	1,000	0.751	751
4	1,000	0.683	683
5	1,000	0.621	621
6	1,000	0.565	565
7	1,000	0.513	513
			$4,868

SOME ECONOMIC INFLUENCES

Table 3-2 Sum-of-years'-digits depreciation

End of year (1)	Depreciation fraction (2)	Depreciation expense (2) × $7,000 = (3)	Present-worth factor at 10% (4)	Present worth (3) × (4) = (5)
0			1.000	
1	7/28	$1,750	0.909	$1,591
2	6/28	1,500	0.826	1,239
3	5/28	1,250	0.751	939
4	4/28	1,000	0.683	683
5	3/28	750	0.621	467
6	2/28	500	0.565	283
7	1/28	250	0.513	128
Total	28/28	$7,000		$5,330

$$1 + 2 + 3 + 4 + 5 + 6 + 7 = 28$$

This can easily be calculated by the equation

$$\text{SOYD} = \frac{n(n+1)}{2} = \frac{7 \times 8}{2} = 28 \quad \text{where } n = \text{life in years}$$

Once the sum of years' digits is calculated, the annual depreciation expense is found by multiplying the depreciable amount by the fraction (years of remaining life) divided by (sum of years' digits). This fraction is equal to $\frac{7}{28}$ for the first year, $\frac{6}{28}$ for the second year, $\frac{5}{28}$ for the third year, etc.

Table 3-2 shows the computations for the present worth of sum-of-years'-digits depreciation for the illustrative example. At 10 percent interest, the present worth of the depreciation expenses is $5,330.

Double-declining-balance depreciation, the other approved method of accelerated depreciation, is calculated as follows. The annual percentage rate is determined by dividing 200 percent by the number of years of useful service life. In the example problem above, this is (200%)/7 = 28.6%. The annual depreciation expense is then determined by multiplying the percentage rate times the book value of the property at the *beginning* of the tax year. This depreciation expense is subtracted from the book value at the beginning of the year to determine the book value at the *end* of the year. Note that no salvage value is used in this computation, only the book value. This will result in a rapid depreciation approaching zero value, so that the law clarifies this procedure by saying that an item shall not be depreciated below a reasonable market value. Assume in this case that the reasonable market value is the estimated salvage value of $3,000.

Table 3-3 Double-declining-balance depreciation

End of year (1)	Book value (2)	Depreciation expense 0.286 × (2) = (3)	Present-worth factor at 10% (4)	Present worth (3) × (4) = (5)
0	$10,000		1.000	
1	7,140	$2,860	0.909	$2,600
2	5,098	2,042	0.826	1,687
3	3,640	1,458	0.751	1,095
4	3,000	640	0.683	437
Total		$7,000		$5,819

Table 3-3 gives the computations for this type of depreciation, showing a present worth of $5,819 at 10 percent interest. Note that the full depreciation expense allowable in the fourth year was not taken, only the amount required to reduce the book value to $3,000. Note that an easy way to determine each succeeding year's depreciation expense is to obtain the first year's expense by multiplying the original cost (book value at time 0) by the annual rate (0.286 × $10,000). After that each succeeding year's expense will be equal to (100 percent − annual rate) times the prior year's expense. The expense for the second year then is 0.714 × $2,860 = $2,042, etc.

When an item does not meet the tests for an accelerated method of depreciation, generally because it is not new, 150 percent declining-balance depreciation often is acceptable. For comparative purposes, assume that for the example machine double-declining-balance depreciation cannot be used. In this case the annual percentage rate is determined by dividing 150 percent by the number of years of useful service life. This annual rate is (150%)/7 = 21.4%. Table 3-4 shows the calculations for the

Table 3-4 150% declining-balance depreciation

End of year (1)	Book value (2)	Depreciation expense 0.214 × (2) = (3)	Present-worth factor at 10% (4)	Present worth (5)
0	$10,000		1.000	
1	7,860	$2,140	0.909	$1,945
2	6,178	1,680	0.826	1,389
3	4,856	1,322	0.751	993
4	3,817	1,039	0.683	710
5	3,000	817	0.621	507
		$7,000		$5,544

present worth of 150 percent declining-balance depreciation at 10 percent interest. The result is $5,544.

Although expensing is not permitted for equipment which has a useful service life of several years, this will be calculated so it can be compared with several depreciation methods. In this case, it will be assumed that the machine is expensed in the year of purchase and sold for $3,000 at the end of a 7-year life. The $3,000 is treated as a gain on sale and discounted back to the present. Table 3-5 shows that this procedure yields a present worth of $8,460, higher than any of the depreciation methods.

When these methods are ranked according to the present worth of the depreciation, the following result is obtained:

Expensed	$8,461
DDB	5,819
1½ DB	5,544
SOYD	5,330
SL	4,868

This is the usual result of comparing the various methods in question. It is always true with discounting that the quicker you get your money back, the more it is worth. Expensing will always result in the highest return and straight-line depreciation in the lowest. Generally, double-declining-balance depreciation will show a better return than sum of years' digits. However, it is sometimes possible to get a better write-off with sum of years' digits by using a long life and low salvage value. These comments relate to the use of these methods in economy studies and for tax purposes. Other factors influence the choice for costing, estimating, and accounting.

These conclusions are supported by curves[1] (Fig. 3-2) which

[1] James Douglas, "Obsolescence as a Factor in the Depreciation of Equipment," *Stanford Univ. Constr. Inst. Tech. Rep.* 22, 1963.

Table 3-5 Expensed equipment

End of year (1)	Amount expensed (2)	Present-worth factor at 10% (3)	Present worth (4)
0	$10,000	1.000	$10,000
7	− 3,000	0.513	− 1,539
Total	$ 7,000		$ 8,461

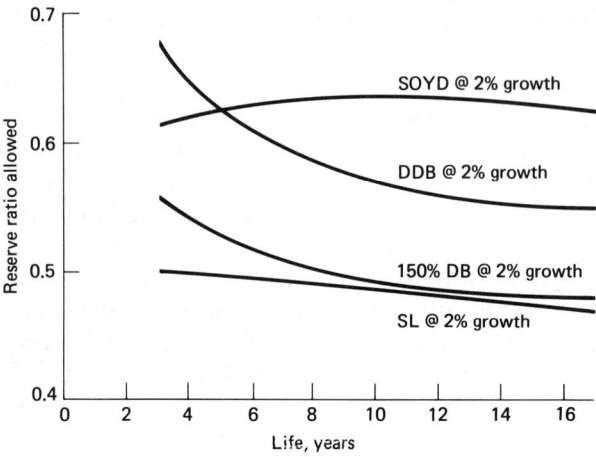

Fig. 3-2 Allowable depreciation reserve ratio.

show the allowable depreciation reserve ratios permitted by Revenue Procedure 62-21. The curves were drawn by using the "Tables for Applying Revenue Procedure 62-21"[1] as a source. Although the reserve-ratio test is no longer required, the tables are indicative of the depreciation reserve which will be accumulated under given conditions. The curves in Fig. 3-2 are derived from data for + 2 percent annual growth in equipment investment. It is clear from this that accelerated methods are superior to straight line; that for short lives of less than 5 years, double declining balance is better than sum of years' digits; and that 150 percent declining balance is better than straight line. Hence it can be concluded that one should generally use double-declining-balance depreciation for new construction equipment with a useful life of more than 3 years and 150 percent declining balance for used equipment.

USEFUL SERVICE LIFE

The useful service life of a machine has long been the subject of controversy in depreciation accounting. It must be assumed when the machine is purchased and must be reasonably consistent with the owner's practice on replacement in the past. Of course, the life assumed has a material effect on the rate at

---

[1] U.S. Treasury Dept., "Tables for Applying Revenue Procedure 62-21," 1962.

SOME ECONOMIC INFLUENCES

which the capital value of the machine is repossessed by the owner. Table 3-6 illustrates the differences in present worth of depreciation expense for useful service lives of 3, 5, and 8 years when the machine was actually retained for 5 years. Assume that a machine was purchased for $10,000 and sold 5 years later for $3,000. Depreciate it by double declining balance and discount at a rate of 10 percent. Assume that any excess depreciation is recaptured by gain on sale at the end of 5 years.

Obviously the maximum present worth is achieved when a depreciation life of 3 years is used, for the money is returned faster. For consistency with the actual acquisition and disposition of the machine as evidenced by the contractor's records, he should use a life of 5 years. In any case it would be disadvantageous to use the longer life of 8 years. By legal interpretation, the useful service life of a machine begins when it is placed in service and ends when it is retired.[1] Retirement has been interpreted to mean disposal by sale or trade and not by placing equipment in the yard on a standby basis.

Bulletin F[2], originally issued in 1926, was used for many years to determine useful service life. Each machine was listed with the established life. As obsolescence brought pressure on owners to get rid of equipment sooner, there was much dissatisfaction over the lives listed in *Bulletin* F. Typical of these lives were:

Equipment	*Life, years*
Dump trucks (2 yd^3 plus)	8
Crawler tractors (20 tons plus)	10
Steam shovels (2 yd^3 plus)	10
Scrapers, carryall	6

As time went on, there was much pressure brought on Congress to change the law. The revenue Code of 1954 brought this change and allowed accelerated depreciation methods (double declining balance and sum of years' digits) for the first time while continuing the use of straight line and 150 percent declining balance. *Bulletin* F was continued at that time for use in establishing

---

[1] U.S. Treasury Dept., "Regulations Relating to Depreciation," *Treas. Decis.* 6182, *Tech. Pap.* 311, pp. 2 and 10, 1957.
[2] U.S. Treasury Dept., "Tables of Useful Lives of Depreciable Property," *Bull.* F, 1942.

Table 3-6 Depreciation expense at various lives

End of year (1)	3 year DDB depreciation (2)	5 year DDB depreciation (3)	8 year DDB depreciation (4)	Present-worth factor at 10% (5)	Present worth 3 year (6) = (2) × (5)	Present worth 5 year (7) = (3) × (5)	Present worth 8 year (8) = (4) × (5)
1	6,667	4,000	2,500	0.909	6,060	3,636	2,273
2	2,222	2,400	1,875	0.826	1,835	1,982	1,549
3	741	1,440	1,406	0.751	556	1,081	1,056
4	247	864	1,055	0.683	160	590	721
5	82	518	791	0.621	51	322	491
Gain	−2,959	−2,222	−627	0.621	−1,838	−1,380	−389
Total					6,833	6,231	5,702

useful service lives. After much agitation and several years of hearings, by Congressional committees, *Revenue Procedure* 62-21[1] was issued by the Treasury Department. This procedure established a guideline life of 5 years for construction equipment. The taxpayer was allowed to adopt any life, however, that would accumulate a depreciation reserve which fell within the allowable limits of the published tables.[2] A reserve-ratio test was described to determine whether the adopted life was consistent with the owner's policy. The latest government policy on life was promulgated in *Treasury Decision* 7128[3] in July 1971. This procedure further modified the guideline life of *Revenue Procedure* 62-21 to permit an asset depreciation range (ADR) which would be 20 percent less to 20 percent greater than the guideline life. For construction equipment, this means that a life of 4 to 6 years can be used without challenge since the reserve-ratio test was set aside for users of ADR. Also established was a maximum annual repair allowance which could be expensed. For construction equipment, this amounted to $12\frac{1}{2}$ percent of the average book value during the tax year.

In summary, it can be said that for depreciation purposes it behooves the owner to keep the shortest useful service life that he can defend reasonably. In many cases, this will be the 4-year minimum life specified in the ADR for construction equipment. A 3-year life is the best you can do and still utilize an accelerated method of depreciation. In any case, the law does not allow depreciation below a reasonable salvage value.[4] Since it is impractical for the engineering revenue agents in the field to enforce this provision of the law, they must rely on the recapture clause which returns gain on sale at the full tax rate. This results in a small amount of deferred interest in favor of the taxpayer but is not a real point at issue. It should be mentioned, however, that the owner has effectively borrowed any gain on sale for some period free of interest.

DETERMINATION OF SALVAGE VALUE

The salvage value of a machine is its reasonable market value at some future date of disposal. It may include a dismantling

---

[1] U.S. Treasury Dept., *Depreciation Guidel. Rules, Revenue Proc.* 62-21, 1962.
[2] U.S. Treasury Dept., "Tables for Applying Revenue Proc. 62-21," 1962.
[3] U.S. Treasury Dept., *Treas. Decis.* 7128, 1971.
[4] U.S. Treasury Dept., Regulations Relating to Depreciation, *Treas. Decis.* 6182, p. 2, 1957.

charge or cost to remove, if applicable. Past practice is the best guide for establishing this salvage value, and the contractor's previous experience is generally given some weight in the determination. If the new ADR technique is used, *Treasury Decision* 7128 permits the reduction of the gross salvage value of a vintage account (group account of assets placed in service in a given tax year) by a maximum of 10 percent of the original cost. For example, if the estimated salvage value of a $1,000 machine is $120 at the end of a 10-year life, the salvage value of $20 [$120 − (0.10 × $1,000)] may be used.

THE INFLUENCE OF INVESTMENT CREDIT

Another tax influence is investment credit. It has been repealed and restored several times, evidently being used by government economists as an economic manipulator. The real purpose of investment credit is to encourage investment by owners in replacement and additional capital equipment. It is not a refund; it is a tax credit applied against income tax due. The amount of credit earned on an item is equal to 7 percent of the invested amount (total cost of a new acquisition, difference between cost of new machine and salvage value of old machine if a replacement). The full 7 percent can be taken only on a machine which has a life of more than 7 years. Two-thirds of that amount may be taken on a machine which has a life of 5 to 7 years and one-third on an item which has a life of 3 to 5 years.

The law provides for carry-back and carry-forward of any unused amounts of investment credit. This means that it can be applied to back taxes for the previous 3 years and future taxes for the next 7 years. The maximum amount of investment credit allowed is equal to the tax due if less than $25,000. If the tax due is more than $25,000, the maximum is equal to $25,000 plus one-half the difference between tax due and $25,000. For example, if the tax due is $45,000, the maximum credit allowed would be

$$\$25,000 + \tfrac{1}{2}(\$45,000 - \$25,000) = \$35,000$$

As a practical matter, it is difficult to predict accurately the useful service life of a machine when it is purchased. If the useful service life of a crawler tractor is estimated to be 5 years, then two-thirds of the 7 percent investment credit might be taken in the year of purchase. If the tractor is sold at the end of 3 years, one-half the credit taken must be returned to the government. As one contractor's comptroller put it: Where else can I borrow that amount of money with no interest?

Whatever is claimed for investment credit as an economic stimulus, it cannot be said to be uniform in its effect on construction equipment. In our present era of technological advancement, much equipment will have a life of less than 4 years. Most of it will have a life of less than 5 years. It is likely then that some will be eligible for one-third of the 7 percent credit, and only a small amount will be eligible for more than that. Under present equipment policies, it is unlikely that the law as presently written will have much effect on equipment economics.

THE RECAPTURE OF INCOME

Before Jan. 1, 1968, gain on sale was taxed as a long-term capital gain at 50 percent of the tax rate. There was a great effort on the part of contractors to get a fast depreciation rate in order to take advantage of this favorable procedure. Arguments between owners and tax authorities were legion because of the large amount of money involved. Since the inclusion of the recapture clause in the tax code, most arguments have ceased. The recapture clause states that separate depreciation accounts (item accounts) must be kept on each machine, and at the time of disposal, the full tax will be paid on the gain on sale. The result of fast write-off now is no more than some interest on the recaptured funds. This has had a very quieting effect on collection of taxes and has reduced the friction previously evident.

INFLATION AS AN ECONOMIC FACTOR

Another important economic influence in equipment analysis is inflation. For the past several years inflation has placed increasing pressure on our economic system. The total effect of inflation is complex when combined with other factors which affect the economics of equipment.

Inflation is the result of an excess of money and available credit over the supply of goods. This causes the price of goods to increase. When there is a continuous rise in prices sustained by a continuing increase in wages, the two act upon each other to cause an inflationary spiral.

Inflation behaves just like compound interest. It takes more dollars of future years to buy an equal amount of a given com-

Table 3-7 Present value of inflated dollars

Year	Cost of widget	Present value of future dollar
0	$1.0000	$1.0000
1	1.0500	0.9524
2	1.1025	0.9070
3	1.1576	0.8638
4	1.2155	0.8227
5	1.2763	0.7835

modity, whether a material or a service. The money problems created by inflation result from the combination of dollars of different years when they are not of equal value. This occurs primarily in depreciation accounting, tax computations for gain on sale, and loan interests. The first two work to the disadvantage of the equipment owner while the latter is to his advantage. Let us see why.

Assume that today one can buy a widget for a dollar. If we have an inflation rate of i annually, 1 year from now the widget will cost $(1 + i) \times \$1$, and 2 years from now it will cost $(1 + i)^2 \times \$1$, and so on. Assuming 5 percent annual inflation, Table 3-7 shows the cost of widgets in future years and the present value of the future dollar used to purchase them. The present value of the inflated dollar is found by "recounting" back to the present. This is equal to $\$1/(1 + i)^n$, where i is the annual rate of inflation and n is the number of years hence. This is identical to discounting dollars to obtain the present value from compound-interest formulas. A glance at the interest tables in any good reference book will confirm that these are also discounts at 5 percent compound interest. Money has a time value related to the rate of inflation as well as the interest rate, and the two are additive.

Example A contractor buys a new machine for $12,000. From past experience he decides to use a 5-year life and an estimated salvage value of $2,000 for depreciation purposes. He uses a 10 percent rate of interest for analysis. He would like to know the effect of an increase in inflation from 0 to 5 percent on his depreciation expenses.

Solution Find the present worth of his depreciation expenses as inflation goes from 0 to 5 percent while the 10 percent rate of interest is held constant. Compare straight-line and double-declining-balance depreciation expenses at the two rates. For the purpose of straight-line depreciation, the depreciable amount will be $10,000 ($12,000 − $2,000) or an annual expense of $2,000 for each of 5 years. For double-declining-balance depreciation, the annual depreciation expense will be 40 percent [(200 percent/5 percent)] of the remaining book value each year. In order to avoid

SOME ECONOMIC INFLUENCES

conflict with the Internal Revenue Service, no additional depreciation will be taken when book value reaches $2,000 (estimated salvage value) in the fourth year. Table 3-8 shows the results of these computations.

The discount factors for 10 percent interest are used for zero inflation (10 percent interest + 0 percent inflation) and 15 percent for 5 percent inflation (10 percent interest + 5 percent inflation). The results show that the present worth of these depreciation deductions for both methods are markedly less with 5 percent inflation than without it. In the case of straight-line depreciation, it amounts to $874, while for double declining balance it is $615 less. Contrasting the two methods at zero inflation shows $864 difference in favor of declining balance. At 5 percent inflation the difference increases to $1,123 in favor of declining balance.

It is quite apparent that in times of inflation the difference between double-declining-balance depreciation and straight line is even greater than in time of no inflation. Thus it may be said that the use of accelerated methods of depreciation in times of inflation is more imperative than when there is no inflation.

But how does one measure the rate of inflation? There are many ways, each commodity or mix of commodities and services having its own rate. No reliable index is known to be available to measure the total effect of costs on equipment ownership and operation. The Consumer's Price Index (CPI) is the most common measure and generally is used in union agreements on wages to determine the cost-of-living increases. It is also used by the federal government to determine the value of the dollar to some base year. The present base year for the CPI is 1967 (+ 100). In December 1973 the CPI was 138.5, which is to say that it was 38.5 percent higher than in 1967. Conversely, the 1973 dollar was worth only 72.2 cents ($1/1.385) compared with the 1967 dollar.

The CPI measures the monthly changes in the prices of hundreds of goods and services purchased by consumers in 56 cities across the country. The package is made up of the following proportions of income:

Housing	33.0%
Food	22.5%
Health and recreation	19.5%
Transportation	14.0%
Clothing	10.0%
Miscellaneous	1.0%

Other price indexes measure other mixes of labor and materials. Some of the best known are the Wholesale Price Index (WPI), the Engineering News-Record Cost Indexes (Building and Construction) (ENR), and the Federal Highway Administration

Table 3-8 Present worth of depreciation with inflation

Year (1)	SL depreciation expense (2)	DDB depreciation expense (3)	10% discount factor (4)	15% discount factor (5)	SL at 0% inflation (6) = (2) × (4)	SL at 5% inflation (7) = (2) × (5)	DDB at 0% inflation (8) = (3) × (4)	DDB at 5% inflation (9) = (3) × (5)
1	$2,000	$4,800	0.909	0.870	$1,818	$1,740	$4,363	$4,176
2	2,000	2,880	0.826	0.756	1,652	1,512	2,379	2,177
3	2,000	1,728	0.751	0.658	1,502	1,316	1,298	1,137
4	2,000	592	0.683	0.572	1,366	1,144	404	339
5	2,000	0	0.621	0.497	1,242	994	0	0
Total	$10,000	$10,000			$7,580	$6,706	$8,444	$7,829

$864 $1,123 $615

SOME ECONOMIC INFLUENCES

Table 3-9 Various rates of inflation as of June 30, 1974

	CPI, %*	WPI, %*	ENR, %† Bldg. Const.		FHWA, %† Hwy. bids	Const, %† wages
Last year	11.1	14.5	5.4	5.2	38.0	9.7
Last 4 years	6.1	9.0	9.5	9.7	13.5	9.3
Last 10 years	4.7	6.1	7.0	7.9	7.3	6.2

*Economic Indicators, Congressional Joint Economic Committee and Council of Economic Advisors, United States Government Printing Office, August 1974.
†*Engineering News-Record*, Sept. 19, 1974.

Highway Bid Price Index (FHWA). The various rates of inflation as measured at the end of the fiscal year on June 30, 1974, by these indexes were as shown in Table 3-9.

None of these, not even the CPI, is an exact measure of the rate of inflation for equipment ownership and operating costs. What really is needed is an equipment index which has a proper mix of costs so that an owner can evaluate the inflationary effect on his equipment policy. An index of this sort should contain approximately the following mix of costs:

Machinery price	20%
Prime rate of bank loans	4%
Labor (operating and maintenance)	50%
Parts cost	15%
Fuel, oil, and grease	7%
Overhead (insurance, storage, etc.)	4%
	100%

While one might argue about the relevance of the indexes shown in Table 3-9 to equipment costs, it is abundantly clear that inflation is occurring at a present rate of 5 to 15 percent. It also appears to be increasing, since the rate for the past year is much higher than the average for the past 4 years. Both are considerably higher than the 10-year average. The average annual increase in the CPI from 1913 to 1967 was only 2.25 percent. Truly an inflationary spiral has set in, and it must be considered in framing equipment policy.

A small amount of inflation is characteristic of a healthy American economy. An inspection of source data[1] reveals that a carpenter's hourly wage in 1861 was 6.9 cents. In 1961 his basic wage was $4.30/hour. Over the intervening century, his wage

[1] U.S. Dept. Commerce, "Historical Statistics of the United States, Colonial Times to 1957," 1960.

increased an average of 4.25 percent annually. It would appear that some inflation has been good for us and the nation has responded with an increase in productivity to offset the higher wage rates. This has not been true over the recent years and we have ample cause to worry. The difficult problem that confronts us is how to account for these changes in our equipment policy. The solution to this problem will be discussed later in connection with the use of a mathematical model to determine the economic life of equipment. Figure 2-1 shows the results of a study[1] by use of this model, already discussed in the preceding chapter. With an increase in inflation, profits are depressed and economic life is lengthened. It is sufficient to say that inflation is a problem which has increased with time and now exerts considerable influence on the economic decisions which must be made on equipment ownership.

OBSOLESCENCE AS A FACTOR IN DEPRECIATION OF EQUIPMENT

Obsolescence, like inflation, has come to exert considerable influence in the construction-equipment business. Depreciation is the decline in the capital value of an asset as it ages or is expended in use. Both physical deterioration and economic decline are a part of this aging and wearing process. The economic-decline part is called obsolescence. A more complete treatment of the subject will be found in *Technical Report* 22 of the Stanford Construction Institute.[2]

It follows that depreciation has two constituents which are virtually independent of each other in their effect on earnings; one is physical, and the other is economic. We have been aware of physical deterioration for many years. The literature of economics indicates clearly that depreciation was invented to take care of the decrease in value incident to physical expenditure of the asset. In the modern world technological progress is so rapid that the competition of improved new models has forced the old equipment into economic decline.

We have already discussed the rapid technological changes of the last two decades. These improvements have created an economic inferiority of the old machine to its improved replacement, i.e., obsolescence.

[1] James Douglas, The Influence of Inflation on the Economic Life and Profits of Equipment, *AIME Annu. Meet., San Francisco, February 1972.*

[2] James Douglas, "Obsolescence as a Factor in the Depreciation of Construction Equipment," 22, *Stanford Univ. Constr. Inst. Tech. Rep.* 22, May 1963.

Obsolescence can affect the economics of a machine through several agents. One is the loss in revenue (or revenue foregone) when the present machine is threatened by an existing new model which has higher productivity. The denial of this additional revenue by failure to replace now results in less revenue to the owner. As the present machine continues to age, the revenue-earning gap widens because newer machines with higher and higher earning capacity are made available.

Another agent affecting machine economics is the reduced cost of maintenance and operation due to improvements in replacement machines. It is easier to recall the breakthroughs in machine design than the continuous small improvements. Yet it is continuous improvement in small things like metallurgy in wear parts of engines, bonded brake linings, plastic insulation of electric wiring, pressurized cooling systems, and so on that create continuing obsolescence. When a breakthrough occurs, like the introduction of the powershift transmission, a large amount of obsolescence occurs at one time. This is the exception rather than the rule, and it is wiser to think of obsolescence occurring at a constant rate since there is no way to forecast the breakthrough technology.

Obsolescence, then, is the economic inferiority created in an old machine by the introduction of an improved model. It tends to improve revenues and reduce costs as newer and better machines are made available. As a result of employing improved equipment, the contractor-owner is able to make more profit. The effect of obsolescence on the equipment is to reduce its economic life, i.e., the life which maximizes its profit potential. This will be discussed in greater detail in a later chapter. The greatest effect of obsolescence is felt when the rate of technological improvement in a machine is greatest. It is safe to say that the rate is not yet leveling off. It is a fair prediction that when the full effect of space technology is felt in the construction industry, obsolescence may accrue at an even faster rate than it has during the past decade. Space metallurgy, power generation, electronics (especially solid state and microminiaturization), automation, and other new ideas will continue to modify the construction processes and equipment. Economic life is constantly being shortened by new technology.

IMPORTANCE OF RECOGNITION OF ECONOMIC FACTORS

The economic factors treated in this chapter have a great effect on the decisions which must be made about equipment. Of paramount importance are the answers to such questions as

1. What machine shall I buy?
2. How long shall I plan to keep it?
3. When shall I actually replace it?

The answers to these questions are intimately related to such factors as how much taxes will be paid and when, the prevailing obsolescence rates of a particular machine or class of machines, what rate of inflation will exist during the life of the machine and its successors, and what other economic factors may influence these decisions.

In the modern competitive environment, to ignore these economic factors is to deny them. It is far better to make a reasonable estimate of an unknown factor than it is to deny it—which is to equate its effect to zero. What is needed is a method of analysis which can account for these factors. If their true values cannot be known, there are ways of determining the degree of influence so they may be properly weighted in making decisions where they are involved. Such a method of analysis will be developed in the next chapter concerning the optimum life of equipment.

4
Life of Equipment

Some of the most important decisions about equipment are related to its life. These are really two sets of related problems, one set being concerned with the present machine and the other set with future replacements. The first set will be called the replacement problem and the second set the economic-life problem.

Replacement is a problem which has plagued the equipment owner since the advent of the Industrial Revolution. More particularly, the last two decades of expanding technology have made the problem more urgent. Before the Industrial Revolution, the rate of technological improvement in machines, as with most durable goods, was very slow. They were generally replaced when they wore out or their physical life ended; i.e., they were no longer able to produce. Life was serene, and it was hardly necessary to use complex analysis to determine whether or not they should be replaced.

Today the situation is entirely different. Modern technology has created obsolescence, and this factor has shortened the economic life of a machine so that it is usually less than its physical

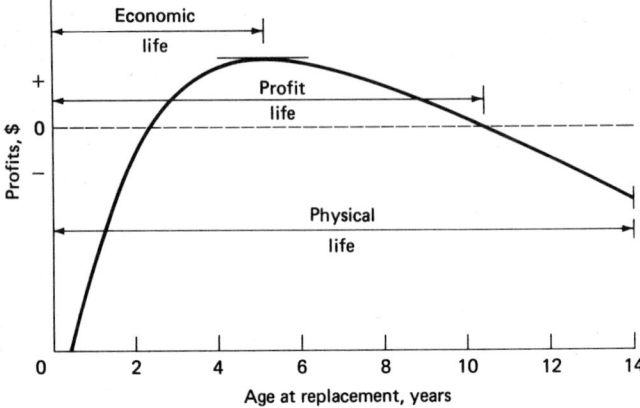

Fig. 4-1 Profits versus age at replacement.

life.[1] Consider these definitions as illustrated in Fig. 4-1, where total profits are plotted versus age at replacement for a series of future machines.

The physical life is the age at which the machine is exhausted and can no longer produce. At this point it will usually be abandoned or scrapped. Profit life is the life over which it can earn a profit; retention beyond that point will create a loss. The economic life of a machine is the life which maximizes profits over the lifetime of the enterprise.[2] It is easy to understand why a satisfied owner is reluctant to replace a machine which has performed dependably in the past and which may still be earning a profit for him. Sentiment must not interfere with good business judgment, however. A glance at Fig. 4-1 shows that this machine has the ability to earn a profit if replaced at 10-year intervals. It will earn many times that profit if replaced at 5-year intervals, the economic life of the machine. This is the basis of the solution of the economic-life problem.[3]

PHYSICAL LIFE

As a piece of construction equipment ages in use, the maintenance- and operating-cost curves rise. The rate of rise depends

---

[1]James Douglas, Obsolescence as a Factor in the Depreciation of Construction Equipment, *Stanford Univ. Constr. Inst. Tech. Rep.* 22, 1963.

[2]James Douglas, Construction Equipment Policy: The Economic Life of Equipment, *Stanford Univ. Constr. Inst. Tech. Rep.* 61, 1966.

[3]James Douglas, Optimum Life of Equipment for Maximum Profit, *J. Constr. Div. ASCE*, January 1968.

LIFE OF EQUIPMENT

on the quality of the machine, the care it receives in use, the nature of the job it is doing, and the quality and amount of maintenance it receives. One of the biggest problems in getting reliable costs is that many owners simply do not have an adequate cost-record system to do the proper job. Another difficulty in obtaining reliable costs is that contractors do not like to reveal them. In spite of these difficulties it is a well-known fact that these costs rise as the machine deteriorates physically.

As indicated, it is an established fact that these costs do increase, generally in proportion to the hours of operation. While it is true that certain machine elements such as gaskets, seals, rubber items, and organic materials usually deteriorate with age, the *wear* on the "iron" is the primary determining factor in the physical life of construction equipment. These costs of wear and tear may be thought of as variable costs, for they vary with the hours of operation.

ECONOMIC LIFE

The net earnings of a machine (revenue minus costs) decline with age and hours of operation. Both quality and quantity of product decline as physical deterioration takes place, which results in a drop in revenue. But the chief influence on revenue is the competition of improved models which can do the job at a cheaper price. It is through the interrelation of costs and revenue that obsolescence manifests itself. Obsolescence is independent of hours of operation and varies with the age of the machine. That portion of the economic decline may be thought of as a fixed cost, for it has a time variance (rather than hours of operation).

As new, improved models of a machine come on the market, economic pressure is brought on the old machine. It is increasingly difficult for the old machine to show a profit in the face of a twofold economic decline, part of the decline due to aging and part due to increased competition from newer machines. Finally, the old machine can no longer generate a profit in its present commitment. Its profit life on this assignment is over. It must be replaced in this function if the owner is to continue to make a profit. This he will do by assigning a new (for that job) machine and either downgrading the assignment or disposing of the old machine. It should be understood that as an alternative, the old machine may be modified and improved to provide a different machine for the job.

Thus, it is evident that a machine has two separate and distinct lives: a physical life and an economic life. The physical life of a machine is the lifetime over which it is physically able to produce, but not necessarily at a profit. The economic life is the lifetime over which it can profitably survive in the face of improved models, changes in methods and techniques, and other variable economic influences. These lives are related but exhibit different characteristics. Before the twentieth century, machines were usually worn out before replacement. Technological improvements came along so slowly that sound policy dictated the total expenditure of a machine. During this era, the physical life governed the replacement decision.

More recently, as the pace of the technological revolution has increased, economic life has become the governing factor. The first time this lesson was really driven home was after World War II, when diesel locomotives replaced steam on railroads in the United States. In this case, which had national attention, steam locomotives were retired long before physical deterioration was complete. The same thing has happened in air transport, where the jet plane has replaced the propeller type, and in the construction industry, where the rotary air compressor has replaced the reciprocating compressor.

REPLACEMENT LIFE

The replacement problem is related to the replacement of an existing machine which has already been partially expended in service. The replacement problem is one of timing (*when* do you replace?) and selection (*what* machine?). The replacement life is defined as the remaining life of the present machine (the one now owned) which optimizes the advantage to the owner. The economics of ownership of the present machine must be compared with the economics of ownership of a series of future machines which could replace it. The solution for the economic life of future replacements should be worked out first for each model of machine considered a potential replacement. Next the replacement life of the present machine must be found. This will be illustrated by simplified examples to show how it can be accomplished.

There are at least three solutions to the problem:

1. Intuition
2. Minimizing costs
3. Maximizing profits

LIFE OF EQUIPMENT

Intuition is without a doubt the most popular method of making replacement decisions. Some call it mature judgment; some say they have a "feel" for these things. However you explain it, there must be a better way. Machines are often replaced when they require a major overhaul and sometimes when a new job is started. At other times, availability of capital is a determining factor because no reserve has been built up in anticipation of replacement. None of these judgmental decisions has a sound economic basis to be used as a criterion for an orderly, planned replacement program.

Second to intuition in popularity is the objective of minimizing equipment costs. Since all contractors keep some records of these costs, this is an easy way out and considered by some to be more scientific than the method described above. The largest deficiency of this minimum-costs method is that all costs are seldom considered, even by estimation. Penalty costs for downtime, obsolescence costs, labor costs, tax expense (consideration of depreciation methods available), and inflation are often ignored.

One equipment manager for a large West Coast heavy-construction contractor keeps a plot of the cumulative cost of each of his large machines. It is his strong opinion that when these costs become excessive and his plotted line rises sharply, it is time to replace the machine. That this is not logical economic analysis will soon be demonstrated. In the first place, it is too late for a prudent decision to be made after the trend is observable on a chart. By then, the costs are sunk and not recoverable. Proper replacement timing depends on a wise prediction of *future* costs. For example, a completed major overhaul (sunk cost) will affect these future costs: downtime and penalty charges (better availability and reliability), maintenance and operating costs (improved condition of machine), and higher resale value. The greatest weakness in the line of logic followed by this manager is that there is no consideration of the future—either the future cost of the present machine or the future cost (and capabilities) of the replacement machine.

As a matter of interest, let us go deeper into this cost-minimization technique so we can compare it later with profit maximization. Perhaps the best way to do this is by means of a simple illustrative example. Remember that in setting up this example problem, all costs are assumed and not necessarily related to real-life numbers; they are approximate and only used to demonstrate a principle.

AN ILLUSTRATIVE EXAMPLE

Assume that a contractor is quite certain of having contracts of such magnitude that his heavy-equipment inventory policy will not be influenced by the completion date of a current contract. In other words, he has a continuing need for 15-ton-capacity rear-dump trucks. The type A trucks he presently owns cost $17,000 each and are now 1 year old. His records from previous jobs with comparable haul roads indicate that the annual maintenance and operating cost is $8,000 for the first year and increases $1,000 each year. These trucks can earn a revenue of $20,000 the first year, decreasing $500 per year thereafter.

A new type B 15-ton truck costs $19,000. It also has a first-year maintenance and operating cost of $8,000, but because of technological improvements this cost increases only $600 per year thereafter. The revenue earned by these trucks is the same as the type A since they serve the same purpose. Both trucks are to be depreciated by double-declining-balance depreciation assuming a 5-year useful service life.

The questions which must be answered are:

1. What is the economic life of type A trucks?
2. What is the economic life of type B trucks?
3. When should the type A trucks (now 1 year old) be replaced?
4. What should replace them, type A or type B trucks?

SOLUTION BY MINIMIZING COSTS

The information required to develop these answers by cost minimization is developed in Tables 4-1 and 4-2, which show the

Table 4-1 Average annual cumulative costs of type A trucks

End of year (1)	Annual M & O cost (2)	Annual dep. expense (3)	Annual cost (4) = (2) + (3)	Cumulative cost (5) = Σ(4)	Average annual cumulative cost (6) = (5) ÷ (1)
1	$ 8,000	$6,800	$14,800	$14,800	$14,800
2	9,000	4,080	13,080	27,880	13,940
3	10,000	2,448	12,448	40,328	13,443
4	11,000	1,469	12,469	52,797	13,199
5	12,000	881	12,881	65,678	13,136
6	13,000	529	13,529	79,207	13,201
7	14,000	317	14,317	93,524	13,361

average annual cumulative costs of type A and type B trucks, respectively. In Table 4-1 column 1 refers to data at annual intervals through the seventh year of life of type A trucks. Column 2 is the annual maintenance and operating (M & O) cost for the year specified. Column 3 is the annual depreciation expense figured by the double-declining-balance method. For the first year, this expense will be 40 percent times the book value, or $0.40 \times \$17,000 = \$6,800$. Remaining book value will be $\$17,000 - \$6,800 = \$10,200$. Depreciation expense for the second year will be $0.40 \times \$10,200 = \$4,080$, etc. Column 4 is the sum of columns 2 and 3 and is the total cost for a particular year. Column 5 is the cumulative cost, i.e., the sum of all previous annual costs up to the end of a particular year. At the end of the second year, this will be the sum of the annual costs for the first 2 years, $\$14,800 + \$13,080 = \$27,880$, etc. Column 6 is the average annual cumulative cost, i.e., the cumulative cost of column 5 divided by the number of years in column 1.

The economic life of a machine is determined by the year in which the average annual cumulative cost is minimized. This will result in the lowest cost over a long period of time. It will be observed that this occurs at the end of the fifth year in Table 4-1 and results in a minimum of $13,136. The average annual cumulative cost in column 6 is often ignored by considering only the annual costs in column 4. That is what the manager was doing when he said he plotted the cost and watched for an increase. In this example, the lowest annual cost is in the third year ($12,448). If this were used as a criterion for replacement, it would be time to replace at the end of the third year instead of the fifth.

Now, examine Table 4-2 for an analysis of type B trucks. The columns are the same as Table 4-1, but the data are based on type B truck costs. Here the economic life appears to be 8 years, and the minimum average annual cumulative cost is $12,435. Since this is lower than the $13,136 minimum cost of the type A trucks, type B trucks should be purchased as replacements.

The time to replace is determined by whatever strategy will minimize total cost over the lifetime of the enterprise. In most cases, the time to replace will be when the annual cost of the present machine for the next year exceeds the minimum average annual cumulative cost of the replacement. It may happen that the annual cost of the present machine is higher than the comparative cost but will drop in succeeding years. In this case it may be necessary to compute the average annual cumulative cost for a few succeeding years to develop the correct strategy.

Table 4-2 Average annual cumulative costs of type B trucks

End of year (1)	Annual M & O cost (2)	Annual depreciation expense (3)	Annual cost (4) = (2) + (3)	Cumulative cost (5) = Σ(4)	Average annual cumulative cost (6) = (5) ÷ (1)
1	$ 8,000	$7,600	$15,600	$ 15,600	$15,600
2	8,600	4,560	13,160	28,760	14,380
3	9,200	2,736	11,936	40,696	13,565
4	9,800	1,642	11,442	52,138	13,035
5	10,400	985	11,385	63,523	12,705
6	11,000	591	11,591	75,114	12,519
7	11,600	355	11,955	87,069	12,438
8	12,200	213	12,413	99,482	12,435
9	12,800	128	12,928	112,410	12,490
10	13,400	77	13,477	125,887	12,589

The timing of replacement is determined by comparing the annual cost (column 4) of type A trucks in Table 4-1 with the average annual cumulative cost of the type B trucks (column 6) in Table 4-2. In this case, the minimum average annual cumulative cost of type B trucks is $12,435. The type A truck which is now 1 year old will cost $13,080 next year and $12,448 the following year, neither of which is less than type B's $12,435. The time to trade is now!

The four questions originally postulated to make the replacement decision have now been answered—answered by a consideration of equipment costs. Let us now make the same determination with the objective of profit maximization. In this case, we need to consider revenues as well as costs.

SOLUTION BY MAXIMIZING PROFITS

Table 4-3 shows the computations to determine the average annual cumulative profits for type A trucks. Column 1 again refers to data at annual intervals through the seventh year of life of type A trucks. Column 2 shows the annual revenue as it decreases $500 a year from $20,000 the first year. Column 3 is the annual cost from column 4 of Table 4-1. Column 4 is the annual profit and is figured by subtracting the cost in column 3 from the revenue in column 2. Column 5 is the cumulative profit for all years up through a particular year. The cumulative profit at the end of the second year will be $11,620 ($5,200 + $6,420); for the third year it is $18,172 (11,620 + $6,552); etc. Column 6 is the

Table 4-3 Average annual cumulative profits of type A trucks

End of year (1)	Annual revenue (2)	Annual cost (3)	Annual profit (4) = (2) − (3)	Cumulative profit (5) = Σ(4)	Average annual cumulative profit (6) = (5) ÷ (1)
1	$20,000	$14,800	$5,200	$ 5,200	$5,200
2	19,500	13,080	6,420	11,620	5,810
<u>3</u>	19,000	12,448	<u>6,552</u>	18,172	<u>6,057</u>
4	18,500	12,469	6,031	24,203	6,051
5	18,000	12,881	5,119	29,322	5,864
6	17,500	13,529	3,971	33,293	5,549
7	17,000	14,317	2,683	35,976	5,139

average annual cumulative profit, which is the cumulative profit of column 5 divided by the number of years in column 1.

The economic life of a machine is determined by the year in which the average annual cumulative profit is maximized. This will result in a higher profit over a long period of time. It will be noted that the maximum profit in column 6 is $6,057 and occurs the third year of life. Incidentally, the largest annual profit in column 4 occurs the same year. This contrasts with Table 4-1, where the lowest average annual cumulative cost occurred the fifth year and the lowest annual cost occurred in the third.

Table 4-4 is an analysis of profits for the type B trucks for a period of 8 years. The columns are the same as in Table 4-3 except that data are for the type B trucks. The maximum average annual cumulative profit of $6,295 is in the fifth year. Since this is more than the $6,057 of type A trucks in column 6 of Table 4-3, type B trucks are obviously the correct replacements.

Table 4-4 Average annual cumulative profits of type B trucks

End of year (1)	Annual revenue (2)	Annual cost (3)	Annual profit (4) = (2) − (3)	Cumulative profit (5) = Σ(4)	Average annual cumulative profit (6) = (5) ÷ (1)
1	$20,000	$15,600	$4,400	$ 4,400	$4,400
2	19,500	13,160	6,340	10,740	5,370
3	19,000	11,936	7,064	17,804	5,935
4	18,500	11,442	7,058	24,862	6,216
<u>5</u>	18,000	11,385	6,615	31,477	<u>6,295</u>
6	17,500	11,591	5,909	37,386	6,231
7	17,000	11,955	5,045	42,431	6,062
8	16,500	12,413	4,087	46,518	5,815

The time to replace is determined by whatever strategy will maximize total profits over the lifetime of the enterprise. The time to replace will generally be when the next year's annual profits of the present machine fall below the average annual cumulative profit of the replacement. Again it may happen that the next year's annual profits are rising and cross the comparative profit, in which case the average annual profits for succeeding years must be investigated.

Table 4-5 shows the results of a computation for average annual cumulative profits for type A trucks starting at the end of year 1 (now). The timing of replacement here is determined by comparing the annual profits (column 3) of the type A trucks in Table 4-5 with the maximum average annual cumulative profit of the type B trucks (column 6) in Table 4-4. The maximum for type B trucks is $6,295 for a 5-year life. In the next year, the type A truck which is now 1 year old will earn a profit of $6,420 and in the following year $6,552. The third year from now, earnings will be only $6,031, less than the $6,295 which can be earned by type B trucks. Note that even though the average annual cumulative profit for the type A trucks three years from now ($6,334) exceeds the maximum of the type B trucks ($6,295), replacement at the end of the second year will maximize the total profits returned to the owner. Therefore, the owner should plan to replace the type A trucks with type B trucks in 2 years more.

INTUITIVE ANALYSIS

Now let us go back to see how our four basic questions can be answered intuitively, and then we shall compare all the answers to note the differences. It is difficult to estimate the economic life of

Table 4-5 Average annual cumulative profits of type A trucks from end of year 1

End of year (1)	Years from now (2)	Annual profit (3)	Cumulative profit (4) = Σ(3)	Average annual cumulative profit (5) = (4) ÷ (2)
Now 1				
2	1	6,420	6,420	6,420
3	2	6,552	12,972	6,486
4	3	6,031	19,003	6,334
5	4	5,119	24,122	6,031
6	5	3,971	28,093	5,619
7	6	2,683	30,776	5,129

LIFE OF EQUIPMENT

Table 4-6 Comparison of decisions by three methods

Item	Intuition	Min. costs	Max. profits
Optimum life of type A trucks	?	5 years	3 years
Optimum life of type B trucks	?	8 years	5 years
Time to replace	Wait	Now	2 years
Replacement type	A	B	B

a machine by intuition because in doing so you have ignored the tools of economic analysis. Probably the greatest influence on the replacement decision will be the psychological one of physical condition of the machine. If a machine has been maintained reasonably well and has a fresh coat of paint, the estimate of its remaining life will be optimistic. As for replacement of the present machine, it is probable that the contractor would wait a few more years and replace with type A trucks. He would replace with type A because they cost $2,000 less than type B and perform the same work (revenue is the same). More often than not, contractors are trapped into buying a machine because they can save a few dollars on first cost. No thought is really given to future maintenance and operating costs, and these are usually four or five times the cost of the machine. In comparing costs, it is well to remember that it is not the first cost of a machine that really counts; it is the total cost over a lifetime of useful service.

COMPARISON OF METHODS OF ANALYSIS

Table 4-6 is a summary of the decisions reached by three methods in solving the example problem. It will be informative to note the differences and deficiencies. In the first place, estimates of optimum life differ by several years. It really cannot be estimated by intuition. The estimates derived by minimizing cost are 2 and 3 years longer than by profit maximization. Time to replace varies from an indefinite wait to replace now. The minimum cost method signals "replace now," but maximum profit says "wait 2 years." As for type of replacement, one method says replace with A, and the other two indicate B as a replacement.

One must admit that there is a provocative range of choices, and yet a decision must be made. The first choice to be made is which is the best method. All things considered, maximizing profits seems to be a better way to continue in business than to minimize costs. Sometimes you can make much more money by spending a little. Surely, if making a larger investment will result in a higher percentage of profit, that is the way to go. The same thing is generally true with equipment. More money spent for a better machine should yield more profit. And yet, how many contractors use first cost alone to guide them in equipment selection! Profit maximization is the best technique to use for analysis of replacement and life problems in equipment economy. Cost minimization should be used by default only where profits cannot be defined.

DEVELOPMENT OF A MATHEMATICAL MODEL

The biggest obstruction to careful economic analysis in making equipment decisions is the job of doing these time-consuming and tedious computations. It is now time to consider what assistance the computer can provide, that of quickly and cheaply solving a complex mathematical model to determine the economics of equipment ownership.

Over the past few years, a complex model of this sort has been developed and studied at Stanford University at the Construction Institute. The technique of building this model is described in detail in several published reports,[1] but will be described sufficiently here to permit a manual solution. The problem can be solved quickly and efficiently by the method described in the reports mentioned above. It can also be solved manually in less detail by use of the computer-generated tables of Appendix 3. This solution will be described in the chapter which follows.

The model is very simple conceptually and is best described as a discounted-cash-flow model. Revenues and costs are expressed as exponential functions. The latter are subtracted from the former and discounted back to the present time to yield the present worth of profits after taxes. The model becomes more complex as the finer points of equipment economics are examined until we end up with a group of equations that only a computer can handle.

A mathematical model is simply an equation or group of equations portraying a system. In the development of this model, a group of equations is written to simulate the revenues and costs of owning, maintaining, and operating construction equipment. By manipulating the variables in these equations,

[1] See *Stanford Univ. Constr. Inst. Tech. Rep.* 22, 61, 69, and 85.

LIFE OF EQUIPMENT

the decision maker is able to confirm quantitatively the decisions he must make which involve the economics of his equipment.

The development of all this economic theory related to equipment has been in process for almost 50 years. New concepts have been added from time to time, so that we now have a rather sophisticated body of mathematical theory to support what intuition has been telling us all along. Needless to say, in making decisions about equipment policy, our present environment dictates a consideration of at least the following factors:

1. Time value of money
2. Technological advances in equipment (obsolescence)
3. Effect of taxes (depreciation techniques, etc.)
4. Influence of inflation, investment credit, gain on sale
5. Increased cost of borrowing money
6. Continuing replacements in the future
7. Increased cost of future machines
8. Effect of periodic overhaul costs and reduced availability

What to choose as a measure of effectiveness is of considerable importance. All too often the wrong choice leads to suboptimization, which may lead to erroneous conclusions. It is popular to minimize costs in making economic studies of equipment. It is more difficult to maximize profits, but how else can one predict income-tax effects without some kind of profit? Other factors important to revenue are increased productivity (productivity obsolescence), availability of machines (maintenance policy), and deterioration of the machine with age—all of which affect the earning capacity of the present machine and future machines. Hence, the key to the solution of most problems by use of this model is the optimization of profits.

The time value of money is considered by using the present worth of profits for the present machine and all its future replacements to an infinite horizon. The concept of an infinite horizon is that the firm will remain in business and replace its equipment into the indefinite future. The principal costs and revenues which influence decision making will occur in the near and foreseeable future. Assuming an economic life of 5 years, the first three replacements will go out 15 years. As has been pointed out in a previous chapter, the first 15 years at 10 percent discount will account for about 75 percent of the total future costs. This supports the optimization of the present worth of profits to an infinite horizon.

In this model, revenues and costs are considered for all machines, both present and future. These may be classified as follows:

1. Revenues from the service of the machines
2. Maintenance and operating costs, including annual fixed costs, penalties, and overhead
3. Capital costs, including interest on investment, depreciation charges, and interest on borrowed funds
4. Discrete costs such as engine, track, and final drive overhauls
5. Income and corporation taxes, considering depreciation method, recapture of income on sale, and investment credit

In order to write the equations for each of these costs, it is necessary to express them in some mathematical form. One of the most adaptable of curve forms is the exponential function. Decreasing annual costs can be simulated by using the form $y = e^{-ax}$. Figure 4-2 shows a family of curves of this form. The rate of decrease can be varied by increasing the exponent a.

If annual costs are increasing, then the form $y = 1 - e^{-ax}$ may be used. Figure 4-3 shows a family of curves of this form. One particular advantage of using these exponential functions for costs is that they are easily combined mathematically with continuous-discount factors, which are also exponential functions.

The general method of composing the model is as follows. The profits generated by the operation of a machine and its re-

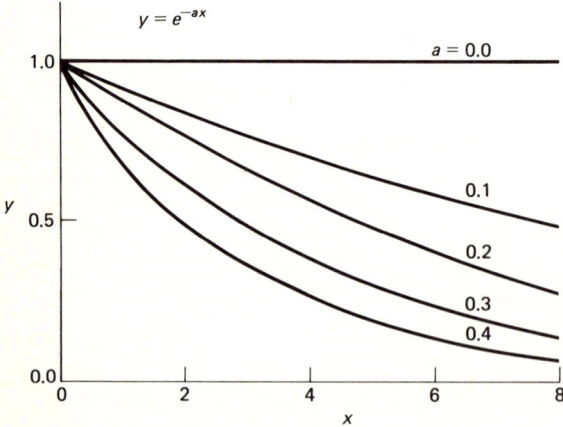

Fig. 4-2 Family of exponential curves $y = e^{-ax}$.

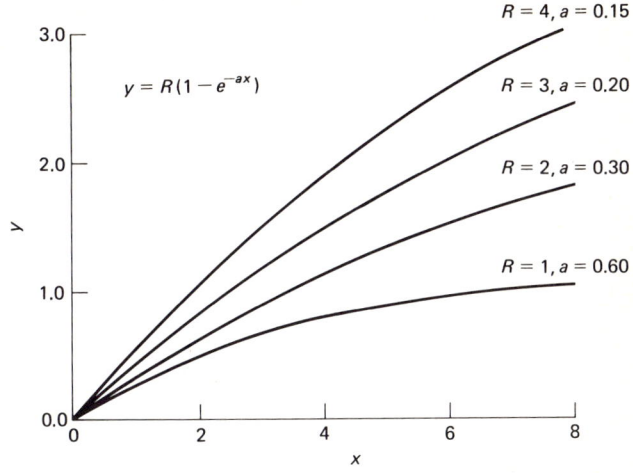

Fig. 4-3 Family of exponential curves $y = 1 - e^{-ax}$.

placements are quantified by writing one or more equations to express each revenue or cost curve. All these dollars are discounted back to the present time, and costs are subtracted from revenues to obtain the present worth of profits after taxes. Quite naturally, taxes are treated as one of the costs of doing business. Variables are listed in Appendix 1 and will be discussed in the ensuing pages in association with the costs which they affect. Very little explanation is required on the curve forms for costs of constant value such as straight-line depreciation expense or redundant costs of overhauls. Their inclusion is simple and direct and will be readily understood when cost curves are discussed. All costs and revenues are reckoned in *present* dollars except those for depreciation, interest charges, and gain on sale. They will be recognized from those curves and equations which contain the inflation factor I_0.

If R_0 equals the present worth of total revenues and E_0 equals the present worth of all expected costs, our objective is to

Maximize $R_0 - E_0$

where

$$R_0 = R_1 + R_2 + R_3$$

and

$$E_0 = E_1 + E_2 + E_3 + E_4 + E_5 + E_6 + E_7 \\ + E_8 + E_9 + E_{10} + E_{11} + I_1 + I_2$$

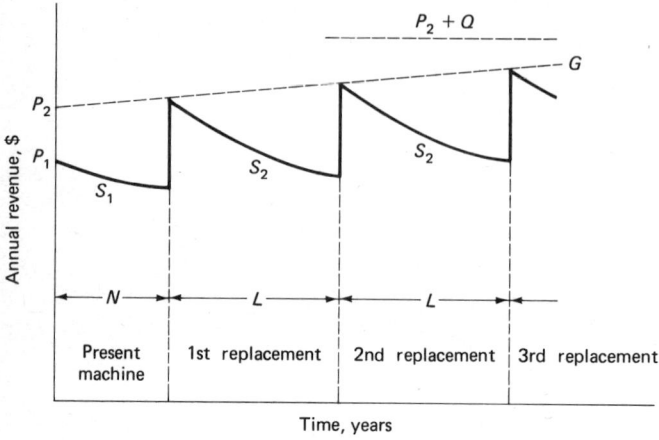

Fig. 4-4 Revenue stream from a machine and its replacements.

Figure 4-4 shows the assumed stream of revenues from a machine and its replacements. P_1 is the present annual revenue of the present machine and P_2 is the present annual revenue of a new replacement. The revenue from a machine is assumed to decline as it ages. Loss of horsepower and vitality, loss of revenue due to breakdown, and the physical deterioration due to wear and tear all contribute to the diminution of the revenue stream as the machine grows older. This rate of decline is given the variable name S_1 for the present machine and S_2 for replacement machines. The productivity of replacement machines increases at the rate G, so that each new machine produces at a higher rate than its predecessor. $P_2 + Q$ is the upper limit of the rise in productivity. This increase in productivity is a reflection of technological improvements which enhance the earning power of machines. Examples of these improvements are increases in capacity and horsepower, automation, and improved designs. N is the remaining life of the present machine, and L is the life of replacement machines. R_6 is the force of interest for continuous compounding.

If R_1 equals the present worth of revenue from present machine and $R_2 + R_3$ is the present worth of revenues from future machines, then

$$R_1 = P_1 \frac{1 - \exp[-(S_1 + R_6)N]}{S_1 + R_6}$$

$$R_2 = \frac{P_2 + Q}{1 - \exp(-R_6 L)} \frac{1 - \exp[-(S_2 + R_6)L]}{S_2 + R_6} \exp(-R_6 N)$$

LIFE OF EQUIPMENT

$$R_3 = -\frac{Q \exp(-GN)}{1 - \exp[-(G+R_6)L]}$$

$$\times \frac{1 - \exp[-(S_2+R_6)L]}{S_2 + R_6} \exp(-R_6 N)$$

The maintenance- and operating-cost equations include all ownership and operating costs not covered elsewhere. Figure 4-5 shows the assumed shape of these curves. A_1 is the present annual cost for the present machine and A_2 is the lower cost of a new replacement. These costs are assumed to rise as the machine ages. This rate is W_1 for the present machine and W_2 for replacement machines. Because of technological improvements which will reduce the cost of maintenance and operation of future machines, the initial cost for each new machine is reduced by the gradient Z. Additionally, the limit of these costs B_2 is reduced by the factor U^{jL}. The cost curves for both revenue and maintenance and operation are reckoned in present dollars. Inflation costs will be discussed with the depreciation curves.

Now let $E_1 + E_2$ equal the present worth of expected maintenance and operating costs of the present machine and $E_3 + E_4 + E_5$ equal the present worth of expected maintenance and operating costs of future machines; then

$$E_1 = (A_1 + B_1)\frac{1 - \exp(-R_6 N)}{R_6}$$

$$E_2 = -B_1 \frac{1 - \exp[-(W_1 + R_6)N]}{W_1 + R_6}$$

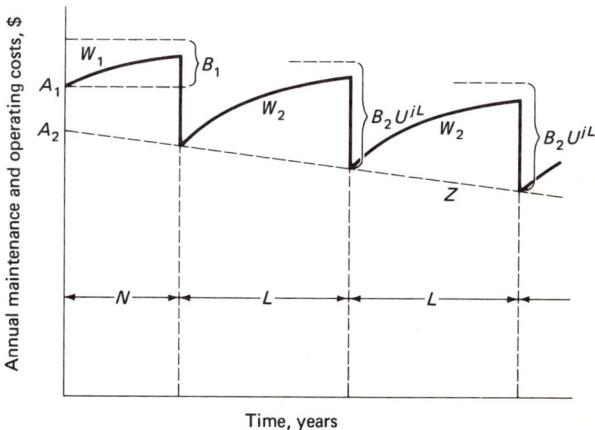

Fig. 4-5 Maintenance and operating costs of machines.

$$E_3 = A_2 \frac{\exp\left[-(Z+R_6)N\right]}{R_6} \frac{1-\exp(-R_6L)}{1-\exp\left[-(Z+R_6)L\right]}$$

$$E_4 = B_2 \frac{\exp(-R_6N)}{R_6} \frac{1-\exp(-R_6L)}{1-U^L\exp(-R_6L)}$$

$$E_5 = -B_2 \frac{\exp(-R_6N)}{R_6+W_2} \frac{1-\exp\left[-(W_2+R_6)L\right]}{1-U^L\exp(-R_6L)}$$

The capital costs which apply to a series of machines are shown in Fig. 4-6. C_1 is the present value of the present machine, which declines at a rate of C_3 until replacement at the end of N years. C_2 is the present cost of a new replacement. The cost of each new replacement machine C_{2j} is assumed to increase at an annual rate equal to V. Since no limit has been placed on V, its rate must be less than the discount factor or the solution is invalid. For a solution where V equals or exceeds R_6, the force of interest for continuous compounding, a new equation with different assumptions must be derived. Immediately after purchase, the value of the machine drops to K times the purchase price and thereafter declines at a rate of C_4. This curve is used to compute the replacement cost and salvage values of all the machines.

Let E_6 equal the present worth of capital costs on the present machine and E_7 equal the present worth of capital costs on all future machines; then

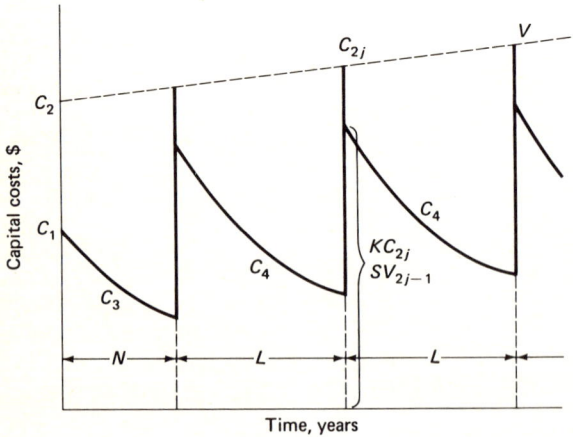

Fig. 4-6 Capital costs of machines.

$$E_6 = C_1 \{1 - \exp[-(C_3 + R_6)N]\}$$

$$E_7 = C_2 \exp[-(R_6 - V)N] \left(1 + \{1 - K \exp[-(C_4 + V)L]\} \right.$$

$$\left. \times \frac{\exp[-(R_6 - V)L]}{1 - \exp[-(R_6 - V)L]} \right)$$

Income taxes paid on the present machine and its replacements are calculated by applying an average tax rate R_5 for federal and state taxes to the difference between revenue and all costs. Items which may be expensed such as maintenance and operating costs, depreciation expense, and interest charges on equipment loans are deducted. Credit for investment is taken where applicable, and any gains on sale are added to revenue. If E_8 equals the present worth of taxes on the present machine and E_9 equals the present worth of taxes on future machines, then

$$E_8 = R_5 [(R_1 + G_1) - (E_1 + E_2 + E_{10} + D_1 + I_1)]$$

$$E_9 = R_5 [(R_2 + R_3 + G_2) - (E_3 + E_4 + E_5 + E_{11} + D_2 + I_2)]$$

$$- IC_1 - IC_2$$

Overhaul costs or any others which occur at periodic intervals are illustrated in Fig. 4-7. The present cost of an overhaul is C_5. These costs are assumed to decrease in each new model because of technological improvements. This decline will take place at the rate C_6. The interval between overhauls is assumed to be a constant, T_3. The time until the next overhaul is due on

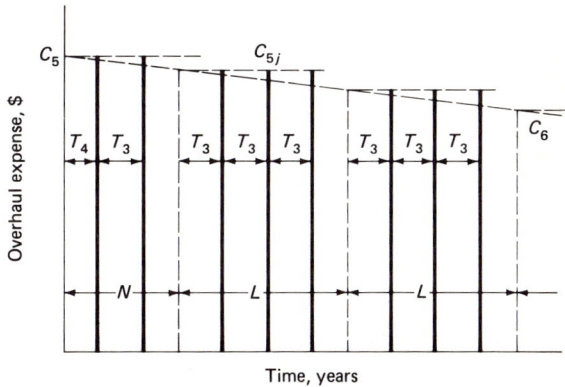

Fig. 4-7 Periodic overhaul costs.

the present machine is T_4. Where several types of periodic overhaul affect the economics of a class of equipment, new variables may be substituted in these equations of discrete costs and the additional equations used in the computations. An example of this would be the periodic engine, track, and final drive overhauls of a crawler tractor. Thus, these equations may be used as many times as needed.

Let E_{10} equal the present worth of all future discrete expenses on the present machine and E_{11} equal the present worth of all future discrete expenses on future machines; then

$$E_{10} = C_5 \exp(-R_6 T_4) \sum_{M_1=0}^{M_1 > (N-T_4)/T_3} \exp(-M_1 R_6 T_3)$$

where $M > T_4$

$$E_{11} = C_5 \exp\{-[C_6 N + R_6(N + T_3)]\} \{1 - \exp[-(C_6 + R_6)L]\}^{-1}$$

$$\sum_{M_2=0}^{M_2 < (L/T_3)-1} \exp(-M_2 R_6 T_3) \qquad \text{where } L > T_3$$

$E_{10} = 0$ where $N \leq T_4$

$E_{11} = 0$ where $L \leq T_3$

Two types of depreciation curves are derived: declining balance (either double or 150 percent) and straight line. Declining-balance depreciation can be demonstrated by Fig. 4-8. Either double declining balance or 150 percent declining balance can be utilized by adjusting the rate of decline, D_3 or D_4. When the sit-

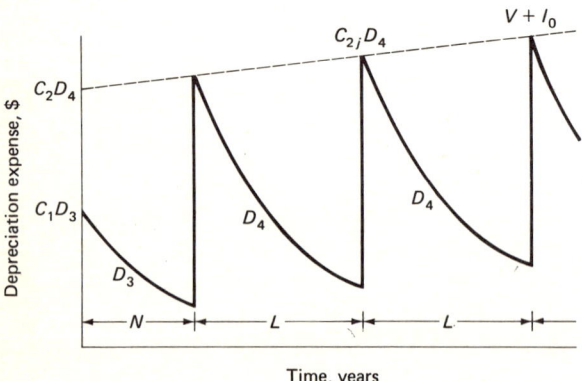

Fig. 4-8 Declining-balance-depreciation expense.

LIFE OF EQUIPMENT

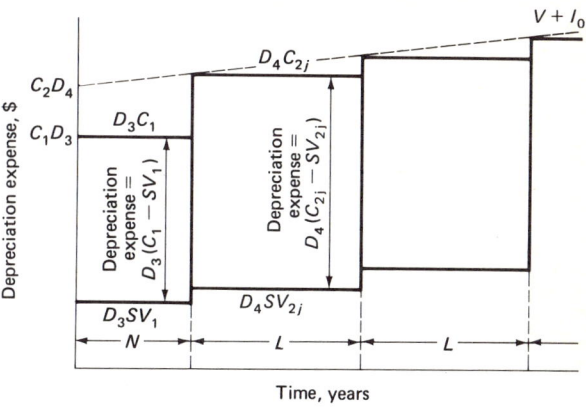

Fig. 4-9 Straight-line-depreciation expense.

uation prohibits using declining-balance depreciation, straight-line depreciation may be used. Figure 4-9 shows how this may be done. In this case, the depreciation expense is equal to the annual rate of straight-line depreciation times the difference between the cost and salvage value of the machine.

Let D_1 = present worth of double-declining-balance depreciation on present machine
D_2 = present worth of double-declining-balance depreciation on future machines
D_3 = coefficient to determine allowable double declining balance on present machine with present age of A_0 years
D_4 = coefficient to determine rate of allowable depreciation on future machines with lives of L years
D_5 = present worth of straight-line depreciation on present machine
D_6 = present worth of straight-line depreciation on future machines

Then

$$D_1 = \frac{C_7 D_3}{D_3 + R_6 + I_0}\{1 - \exp[-(D_3 + R_6 + I_0)N]\}$$

$$D_2 = \frac{C_2 D_4 \exp[-(R_6 - V)N]}{D_4 + R_6 + I_0} \cdot \frac{1 - \exp[-(D_4 + R_6 + I_0)L]}{1 - \exp[-(R_6 - V)L]}$$

$$D_3 = \ln\frac{N + A_0}{N + A_0 - 2} \qquad D_4 = \ln\frac{L}{L - 2}$$

$$D_5 = \frac{C_7}{N+A_0} \frac{1-\exp[-(R_6+I_0)N]}{R_6+I_0} \{1-\exp[(I_0-C_3)N]\}$$

$$D_6 = \frac{C_2}{L} \frac{\exp[-(R_6-V)N]}{R_6+I_0} \frac{1-\exp[-(R_6+I_0)L]}{1-\exp[-(R_6-V)L]} \{1-K\exp[(I_0-C_4)L]\}$$

A few words of advice about inflation are in order. Whenever dollars of different years are combined, as in depreciation economics, a factor must be introduced to account for the decline in the value of the dollar. The variable I_0 is utilized for this purpose. It is used in both Figs. 4-8 and 4-9 to show the increased cost of replacement equipment. The income-tax law equates the high-value dollar at purchase to the lower-value dollar at a later date, thus creating an inequity for owners of depreciable property.

Interest charges on equipment loans are a deductible expense of doing business. As such, they affect the computation of taxes on income. Figure 4-10 shows the method of including add-on interest charges on the equipment loan. Here again, inflation becomes a factor since the dollars returned in amortizing the loan are from a different year from the one in which the principal was borrowed. Interest charges also increase at the same rate as the price increase of new equipment, $V+I_0$.

Now let I_1 equal the present worth of interest charges on equipment loan on present machine and I_2 equal the present worth of interest charges on equipment loans to purchase future machines; then

$$I_1 = \frac{R_4 F C_2}{R_6+I_0} \{1-\exp[-(R_6+I_0)T_1]\} \qquad \text{where } N > T_1$$

$$I_2 = \frac{R_4 F C_2}{R_6+I_0} \exp[-(R_6-V)N] \frac{1-\exp[-(R_6+I_0)T_2]}{1-\exp[-(R_6-V)L]}$$

$$\text{where } L > T_2$$

When $N \leq T_1$,

$$I_1[N \leq T_1] = I_1[T_1 = N]$$

When $L \leq T_2$,

$$I_2[L \leq T_2] = I_2[T_2 = L]$$

Investment credit is a tax credit which is applied in the income-tax computations. It is derived by applying a given percentage to the investment at replacement. The total credit of 7

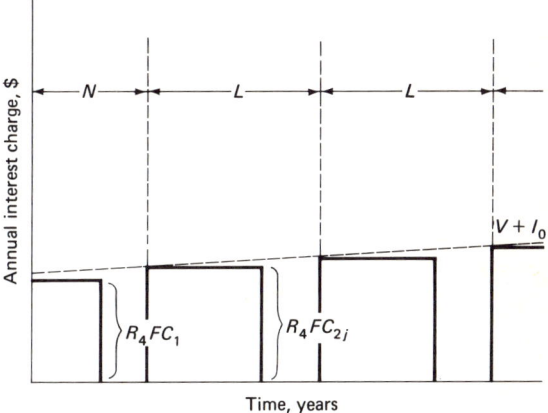

Fig. 4-10 Annual-interest-charge curve.

percent may be used if the machine has a useful service life of 7 years. Two-thirds of this credit is applicable if the machine has a life of 5 years to 7 years and only one-third if the life is 3 years to 5 years. The amount invested is the difference between the cost C_{2j} of the jth replacement and the salvage value of the $(j-1)$st machine, $SV_{2(j-1)}$. These values are derived from the capital cost curves shown in Fig. 4-6.

Let IC_1 equal the present worth of investment credit when present machine is replaced and IC_2 equal the present worth of investment credit for replacement of all future machines.

$$IC_1 = PIC \{C_2 \exp[-(R_6 - V)N] - C_1 \exp[-(C_3 + R_6)N]\}$$

$$IC_2 = PIC \frac{C_2 \exp[-(R_6 - V)N]}{\exp[(R_6 - V)L] - 1} \{1 - K \exp[-(C_4 + V)L]\}$$

Gain on sale occurs whenever book value is less than the salvage value of the machine. This is usually the case with declining-balance depreciation. The equations for straight-line depreciation do not require the use of gain on sale since the depreciable amount is computed in the equation from the capital-cost curve. Gain on sale is applicable only when the useful service life is greater than 3 years and declining-balance depreciation is used. Figure 4-11 shows how Figs. 4-6 and 4-8 are combined to compute these gains.

Let G_1 equal the present worth of gain on sale of the present machine and G_2 equal the present worth of the gain on sale of fu-

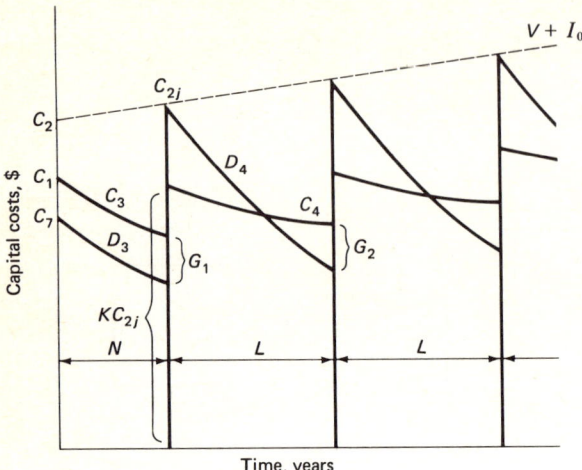

Fig. 4-11 Gain on sale.

ture machines; then

$$G_1 = C_1 \exp\left[(I_0 - C_3)N\right] - C_7 \exp\left(-D_3 N\right)$$

$$G_2 = \frac{C_2 \exp\left[-(R_6 - V)N\right]}{1 - \exp\left[-(R_6 - V)L\right]} \{K \exp\left[-(C_4 + R_6)L\right]$$
$$- \exp\left[-(D_4 + R_6 + I_0)L\right]\}$$

This completes the derivation of the equations presently in use. Equations for other kinds of cost can easily be worked out by the same techniques. Costs may be used flexibly to model almost any situation once the facts are known.

APPLICATION OF THE MODEL

From the curves shown in Figs. 4-4 to 4-11 and the variables listed in Appendix 1, the equations for the model have been derived. These equations are listed in Appendix 2. The solutions to the economic life and replacement problems are found best by use of the computer. However, an approximate manual solution is offered in the next chapter for those who may not be able to take advantage of a computer.

The first step in the solution is to determine the values of the variables. If good equipment-cost records are available over about a 5-year period for the equipment in question, most of the cost and revenue variables can be determined with a fair degree of accuracy. More abstract variables, such as the productivity obsolescence factor G, can be estimated. Suggested values for some of these variables will be found in the next chapter. When all variables have been assigned, a table should be made up listing them.

The second step in the solution is to determine the optimum life of the future machines L, which is the economic life. In setting up the program for the optimization of equipment life, it is necessary to use only those equations which relate to future machines. This can be accomplished easily in the computer program by setting $N = 0$ (which means to replace now). All the present-machine equations will drop out except the one with the salvage value. The value of the present worth of profits after taxes is calculated for the life of future machines at intervals of 0.1 year for about 9.0 years of age. This life span will be adequate for most construction equipment but may be somewhat short for heavy excavators and other heavy or expensive machinery of a special nature. In the latter case, it may be necessary to investigate profits out to a 20- or 30-year replacement cycle. Remember that in letting L vary by increments you are trying different life cycles for a series of replacement machines to an infinite horizon. The value of L that results in maximum profits is the desired economic life of these particular machines. Figure 4-12 shows the result of a computer run to determine the life of a

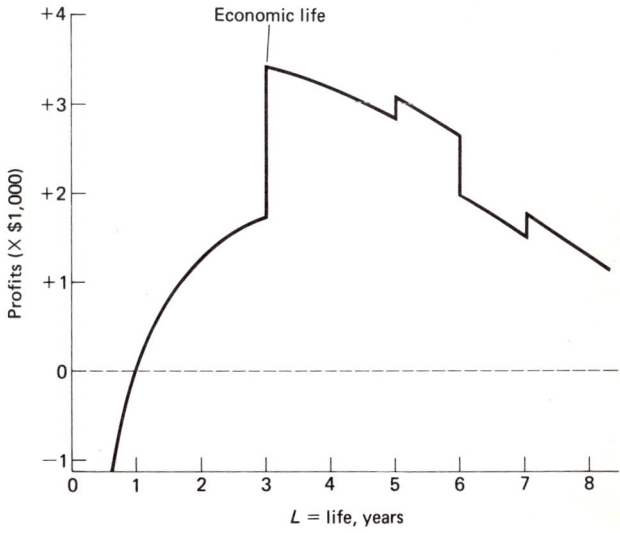

Fig. 4-12 Economic life of a crawler tractor.

typical crawler tractor. In this case, the economic life turns out to be 3 years, the point where depreciation method was changed from straight line to double declining balance. Note also the events of investment credit at 3, 5, and 7 years and periodic overhauls at 3-year intervals. The overhaul at 3 years is blanked by the depreciation change and the investment credit of that year.

The next step in the solution is to determine the replacement life of the present machine. This is accomplished by setting L equal to the economic-life value determined above and iterating N from 0 up to about 5 in 0.1-year intervals. The maximum value of profits in this iteration determines the replacement life (time to replace the present machine). Figure 4-13 is the result of such a study which shows the maximum profit to be at time 0. This signifies that the best time to replace is now. Figure 4-14 shows a study which indicates that the time to replace is about 2.5 years from now.

If several types of replacement machines are to be considered, the economic-life curve of each acceptable replacement must be worked out on the basis of its estimated costs and revenues. Once the type has been determined which yields the maximum profit of all types, it can be used to develop the replace-

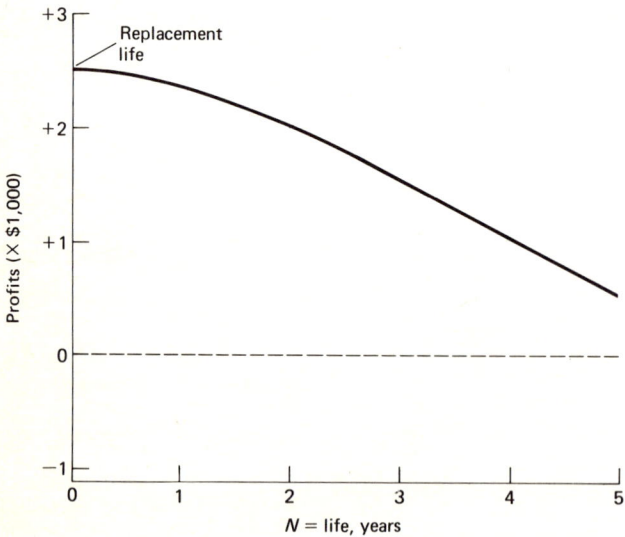

Fig. 4-13 Replacement life-0: replace now.

LIFE OF EQUIPMENT

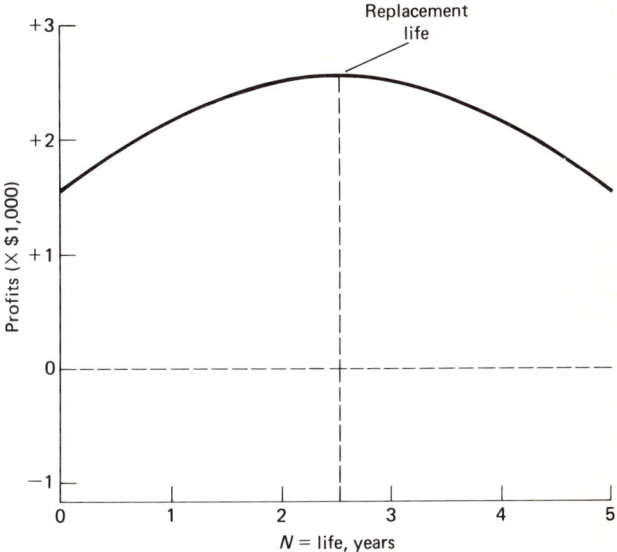

Fig. 4-14 Replacement life-2.5 years: replace 2.5 years from now.

ment-life curve to determine when to replace the present machine. The same steps outlined above will be used to solve the problem manually except that iteration will be at yearly intervals to reduce the amount of work. The manual solutions to typical economic-life and replacement problems follow in the next chapter.

5
Solution for Equipment Life

METHODS

There are three methods for the solution of problems related to economic life and replacement using the model described in the previous chapter. One uses the computer, another is a manual solution using slide rule and adding machine, and the third uses tables. The first two solutions are covered in the various technical reports of the Stanford Construction Institute,[1] and the third method will be explained here.

The computer solution involves the use of the equations described in the previous chapter and listed in Appendix 2. First, a run is made with N set to zero, and L is optimized. The value of L obtained is the trial solution for the economic life of all future replacement machines. Next, a run is made with L set to the trial economic life, and N is optimized. This is the replacement life (time to replacement) of the present machine. On the third run, N is set to the replacement life, and L is again optimized to obtain the economic life when N equals the trial replacement life.

[1] See *Stanford Univ. Constr. Inst. Tech. Rep.* 22, 61, 69, and 85.

SOLUTION FOR EQUIPMENT LIFE 75

A fourth run is then made to optimize N with L set to the economic life obtained in run 3. Since the values of L and N obtained in the third and fourth runs usually are very close to those obtained in the first two, further runs are seldom required. After the final optimization of L and N, the two curves are plotted on a graph for visual inspection and retention. Sample computer output for the example problem will be shown following the problem solution by the tables. The computer is the most efficient and economical method of solving this problem. It takes less than 1 min of execution time on a modern high-speed computer and about 1 min of plotter time.

The life problem can also be solved manually by use of a log-log slide rule and an adding machine. Because of the large number of exponentials in the equations, it is not practical to solve with a manual calculator alone. The problem can also be solved with an electronic calculator which can exponentiate. A slide-rule solution of one single iteration with a given value of L and N takes about 6 to 8 hours. To solve the set of equations the five or six times necessary to get a solution for economic life and replacement life would take 35 to 40 hours. It would take a competent analyst about 2 years (approximately 4,000 man-hours) to duplicate the computer solution. At $5 per man-hour (slave labor!) the complete manual solution would cost about $20,000 in contrast to about $20 for a computer solution.

For the equipment owner who would like a compromise between the computer solution and the manual solution, a solution by use of the tables in Appendix 3 is offered. The tables have been developed on a computer in order to shorten the manual solution. The example problem worked out in the pages which follow was solved with about $2\frac{1}{2}$ hours of computation time using an ordinary desk-top adder with a repeat key for performing the multiplications. An electronic calculator would shorten this computation time somewhat. This solution involved five iterations of the equations and hence compares with the 35 to 40 hours required by the slide-rule solution, a saving of about 90% of the time required.

DEVELOPMENT OF THE TABLES

An examination of the equations in Appendix 2 reveals that each one consists of two quantities, a base variable and complex factor. The base variable is a simple one such as P_1, A_1, B_2, C_2, etc., and the factor is a combination of exponential and table variables. These factors will be found in the tables by reference to the table variables such as S_1, W_2, G, V, etc. The equations are solved by multiplying the base variable by the factor in the table.

DERIVATION OF FACTORS

The basic equations can be written as combinations of a base variable and a factor. For example, Eq. (4) for the revenue R_1 of the present machine may be written

$$R_1 = P_1 r_1$$

where P_1 = base variable
r_1 = factor

In this case, the expression for the factor may be written by inspection as

$$r_1 = \frac{1 - \exp[-(S_1 + R_6)N]}{S_1 + R_6}$$

It is this factor, r_1, which can be found in the table. Its value will be found by reference to the values of N and the table variable S_1. The tables are computed for an equivalent annual rate of interest of 10 percent; i.e., the force of interest R_6 is equal to 0.09531.

Table 5-1 is a list of the equations in Appendix 2 rewritten to show how they will look in the solution of a problem using the tables of Appendix 3. Uppercase letters are the referenced equations and lowercase letters are the factors related to them.

The factors in the tables in Appendix 3 are evaluated for values of N equal to 0, 1, and 2 and values of L equal to 2, 3, ..., 10 and 15. If the values of N and L cannot be optimized within this range, a manual or computer solution is indicated. Ordinarily it is satisfactory to determine economic life and replacement life to

Table 5-1 List of table equations

$R_1 = P_1 r_1$	$E_6 = C_1 e_6$	$I_1 = FC_2 i_1$	
$R_2 = (P_2 + Q)r_2$	$E_7 = C_2 e_7$	$I_2 = FC_2 i_2$	
$R_3 = -Qr_3$	$E_{10} = C_5 e_{10}$	$G_1 = C_1 g_1$	where $C_1 = C_7$
$E_1 = (A_1 + B_1)e_1$	$E_{11} = C_5 e_{11}$	$G_2 = C_2 g_2$	
$E_2 = -B_1 e_2$	$D_1 = C_7 d_1$	$IC_{11} = PIC(C_2 ic_{11})$	
$E_3 = A_2 e_3$	$D_2 = C_2 d_2$	$IC_{12} = -PIC(C_1 ic_{12})$	
$E_4 = B_2 e_4$	$D_5 = C_7 d_5$	$IC_2 = PIC(C_2 ic_2)$	
$E_5 = -B_2 e_5$	$D_6 = C_2 d_6$		

SOLUTION FOR EQUIPMENT LIFE

Table 5-2 List of base and table variables

Equation	Factor	Base variable	Table variable
R_1	r_1	P_1	S_1
R_2	r_2	$P_2 + Q$	S_2
R_3	r_3	$-Q$	G, S_2
E_1	e_1	$A_1 + B_1$	
E_2	e_2	$-B_2$	W_1
E_3	e_3	A_2	
E_4	e_4	B_2	
E_5	e_5	$-B_2$	W_2
E_6	e_6	C_1	C_3
E_7	e_7	C_2	K, V, C_4
E_{10}	e_{10}	C_5	T_3, T_4
E_{11}	e_{11}	C_5	T_3
D_1	d_1	C_7	
D_2	d_2	C_2	V
D_5	d_5	C_7	C_3
D_6	d_6	C_2	K, V, C_4
I_1	i_1	FC_2	
I_2	i_2	FC_2	V
G_1	g_1	C_7	C_3
G_2	g_2	C_2	K, V, C_4
IC_{11}	ic_{11}	C_2	V
IC_{12}	ic_{12}	C_1	C_3
IC_2	ic_2	C_2	K, V, C_4

the nearest year. Any machine which has a replacement life of more than 2 years should be reanalyzed annually until a replacement life is optimized. Once the signal to replace is obtained, the matter usually becomes one of maximizing the salvage value of the present machine and minimizing the cost of the new one, i.e., getting the best possible trade-in.

BASE VARIABLES AND TABLE VARIABLES

Base variables and table variables have already been described: base variables are those to be multiplied by the factor in the tables, and table variables are those required to locate the factor in the tables. The tables are basically organized under the values of N and L. Certain variables which had to be treated as constants in order to reduce the size and complexity of the tables, such as R_6, are not treated as table variables. Table 5-2 is a list of the factors, base variables, and table variables for all the equations. It will be useful in making up a form for the solution of a problem.

ASSUMPTIONS IN COMPUTING TABLES

In order to reduce the tables to a manageable size, certain assumptions were required. These will usually be acceptable. If not, it is suggested that the computer solution be utilized. If the assumptions are reasonably correct, the results will be correct, because you are looking at the differences between profits of various years rather than the absolute values.

Several of the assumptions relate to depreciation method. Two methods are offered: straight line and double declining balance. It is assumed in calculating the double-declining-balance factor for the present machine d_1 that the machine is now 3 years old. The depreciation rate will be 200 percent divided by $N + 3$. This seems logical since a taxpayer is not allowed to use the double-declining-balance method unless the machine has a useful service life of at least 3 years. As a corollary, no factors of d_2 are shown for lives of less than 3 years. Whenever double-declining-balance depreciation is used, there will be a gain (or, negatively, a loss) on sale. This is because no salvage value is used and the machine is depreciated to zero (or, as the law says, not below a reasonable market value). G_1 and G_2 must always be used when D_1 and D_2 are computed. It has been assumed in computing g_1 that the book value of the present machine is equal to the present salvage value; that is, C_7 equals C_1. If this is not true, G_1 can be computed manually.

When straight-line depreciation is used, the machine is depreciated to an estimated salvage value. Since this is calculated on the capital-cost curve, there will never be a gain (or loss) on sale with straight-line depreciation. Therefore, whenever D_5 or D_6 is used, G_1 or G_2 will equal zero and it will not be necessary to compute it.

Investment credit is applied to income tax due, which is calculated in equation E_9. It has been assumed that any credit on the present machine has been taken at the time of purchase and is a sunk cost. It has been further assumed that the correct amount has been taken and that none will be returned to the government at the time of disposal of the present machine. The tables are computed for investment credit for the future without establishing a rate or time. The percentage-investment-credit variable (PIC) takes care of this, as explained later in the example problem. Thus, the analyst can use 7 percent (the present rate) and 3, 5, or 7 years (the present time of investment credit) or whatever the law allows if these are changed.

SOLUTION FOR EQUIPMENT LIFE

The treatment of overhaul costs assumes that they are not accomplished until just after the time they are due. In other words, if you are calculating E_{11} and you expect to overhaul at 3-year intervals ($T_3 = 3$), E_{11} will be zero when $L = 3$. The first overhaul does not show up in the tables until $L = 4$ because it is accomplished a few days after $L = 3$. In the tables, overhaul costs are assumed to decline at an annual rate of 1 percent ($C_6 = 0.01$) because of improved technology. Inflation costs do not appear here because future overhaul costs are estimated in present dollars.

The rate of inflation has been assumed to be a constant 2 percent since this is close to the average value over the past decade. A variation of a few percent will not affect the life of a machine more than a few months. If there is doubt, the computer model should be used to verify results.

The discount rate (or rate of compounding interest) has been assumed to be a 10 percent equivalent rate of interest. This is equal to a force of interest of 9.53 percent ($R_6 = 0.0953$), since $e^{0.0953} = 1.10$.

In the treatment of equipment loans, certain assumptions had to be made about the term of the loans and interest rate of borrowing. It has been assumed that equipment is financed for a term of 4 years at a nominal rate of 10 percent on the full amount of the loan (similar to prevailing loans on automobiles). This rate is subject to fluctuation with changes in economic outlook. It will follow the prime interest rate offered by banks, generally going up with inflation and down with deflation. It has already been assumed that the present machine is 3 years old, and so the remaining term of its loan will be 1 year ($T_1 = 1$). The term of the loan on all future machines is assumed to be 4 years ($T_2 = 4$).

The declining maintenance and operating costs of improved models of the same machine are influenced by two variables, U and Z. U is a reflection of the decreased upper limit of cost on future machines and is assumed to be 0.98. Z is the cost decline which proceeds annually as technological improvements are made. Z is assumed to be 0.01.

RANGE OF TABLE VALUES

The range of variables in the tables in Appendix 3 should be adequate to solve most problems of equipment life common to the construction industry. The tables should also be appropriate for solution of many equipment problems encountered in mining, manufacturing, transportation, and other industries relying heavily on the economic use of vehicles and equipment.

Where exact values of table variables are not listed, interpolation and reasonable extrapolation will give an adequate solution. Most of the values of table variables are in increments of 2 percent, which makes interpolation to an even percentile quite simple. The example below shows a format adapted to the use of an adding machine. Suppose you want the factor r_3 for a value of $S_2 = 0.07$. Assume further that $G = 0.02$ and $L = 8.0$. First, record the values of r_3 for $S_2 = 0.06$ and 0.08. Find the difference by subtraction, halve it, and add that to the value for $S_2 = 0.08$. The result is r_3 for $S_2 = 0.07$:

$S_2 = 0.07$	0.06	7.60220
	0.08	7.13902
	Δ	0.46318
	$\Delta/2$	0.23159
	0.07	7.37061

Occasionally, a double interpolation may be required. Avoid this if possible, for the amount of work is tripled. In the example above, suppose r_3 were desired for $G = 0.03$ and $S_2 = 0.07$. The format is shown below for this interpolation:

(1) $S_2 = 0.07$	0.06	7.60220	(2) $S_2 = 0.07$	6.92645
$G_2 = 0.02$	0.08	7.13902	$G_2 = 0.04$	6.50444
	Δ	0.46318		0.42201
	$\Delta/2$	0.23159		0.21101
	0.07	7.37061		6.71545
(3) $G_2 = 0.03$	0.02	7.37061		
$S_2 = 0.07$	0.04	6.71545		
	Δ	0.65516		
	$\Delta/2$	0.32758		
	0.03	7.04303		

Needless to say, triple interpolations should be avoided at all costs since this is three times as much work as the double interpolation. Assumptions in computing the tables and the selection of table values should obviate this unhappy situation.

The selection of an integral percentile for a table value should be adequate for most analyses. Since we are generally looking for the difference in profits in trying to choose between two replacement machines, the use of a table value of $G = 0.06$, for example, would result in the same choice as the use of $G = 0.063$. It is recommended that integral numbers of percen-

SOLUTION FOR EQUIPMENT LIFE

tile be used when selecting values for C_3, C_4, G, S_1, S_2, V, W_1, and W_2.

Table 5-3 is a list of variable values used in computing the tables of Appendix 3. This table includes those which are ordinarily treated as variables in the computer solution but are treated as constants here in order to simplify the tables.

STATEMENT OF THE PROBLEM

Let us now turn to the solution of an example problem to illustrate the use of the tables. A contractor now owns a 600 ft³/min air compressor for which he has a continuing use. He wants to determine the economic life of an adequate replacement and the time to replace his present machine. Fortunately he has kept excellent records of the costs and use of all his machines for the past few years and is able to construct good cost and revenue curves in order to determine his cost variables.

The first thing which must be done is to accumulate all the cost information required to determine the values of the base and table variables. There are 23 of these, 12 base and 11 table variables. At least three basic curves are required: one for

Table 5-3 List of table-variable values

Variable	No.	Values
C_3	5	0.08, 0.10, 0.12, 0.14, 0.16
C_4	5	0.08, 0.10, 0.12, 0.14, 0.16
C_6	1	0.01 = constant
G	4	0.02, 0.04, 0.06, 0.08
I_0	1	0.02 = constant
K	2	0.70, 0.80
L	10	2, 3, 4, 5, 6, 7, 8, 9, 10, 15
N	3	0, 1, 2
R_4	1	0.10 = constant
R_6	1	0.9531 = constant
S_1	5	0.02, 0.04, 0.06, 0.08, 0.10
S_2	5	0.02, 0.04, 0.06, 0.08, 0.10
T_1	1	1 = constant
T_2	1	4 = constant
T_3	3	2, 3, 4
T_4	2	0, 1
U	1	0.98 = constant
V	3	0.01, 0.02, 0.03
W_1	6	0.02, 0.04, 0.06, 0.08, 0.10, 0.12
W_2	6	0.02, 0.04, 0.06, 0.08, 0.10, 0.12
Z	1	0.01 = constant

Table 5-4 Groups of cost variables

Group	Name	No. of variables	Variables
1	M & O	6	A_1
			A_2
			B_1
			B_2
			W_1
			W_2
2	Revenue	6	G
			P_1
			P_2
			Q
			S_1
			S_2
3	Capital	6	C_1
			C_2
			C_3
			C_4
			K
			V
4	Overhaul	3	T_3
			T_4
5	Loan	1	F
6	Depreciation	1	C_7

maintenance and operating cost, one for revenue, and one for capital cost. Each of these curves has six variables, for a total of 18. The remaining five variables can be obtained without drawing curves. The 23 variables can be grouped as shown in Table 5-4.

CHOOSING VALUES FOR THE VARIABLES

The costs and revenues which are pertinent to problems of this sort are the *future* costs and revenues of the present and future machines. For this reason, equipment analysis is like estimating the cost of doing work in order to bid a job. Historical costs are a necessary and valuable tool for economic analysis, just as they are for bidding. Therefore, all available costs, operating-hour information, availability, utilization, depreciation accounts, and other records should be used in constructing the curves. If annual costs and revenues are erratic because of a highly variable operating schedule, they should be figured on an hourly basis by years and then be reconstructed to give annual costs and revenues on the basis of an average number of scheduled hours per year.

SOLUTION FOR EQUIPMENT LIFE

The easiest way to construct the curves is to set down the values by years. Consider first the construction of the maintenance- and operating-cost curve. Remember that this curve will carry all costs not otherwise included in the other curves of capital costs, overhaul, loans, and depreciation, e.g., storage, overhead, licenses and fees, fines, insurance, and major repairs, as well as the usual costs for operating and maintenance (labor, fuel, oil, lubricants, inspections, routine maintenance, and consumables). Draw a histogram of these costs as shown in Fig. 5-1. These costs are estimates of future costs in present dollars. Do not try to include a factor for inflation, since this will be handled in the tables. Now sketch a curve joining the centers of the bars in the histogram. The intercept of this curve on the vertical axis is the value of A_2, the initial M & O cost of a new replacement purchased now. B_2 is the ceiling approached by this cost as it

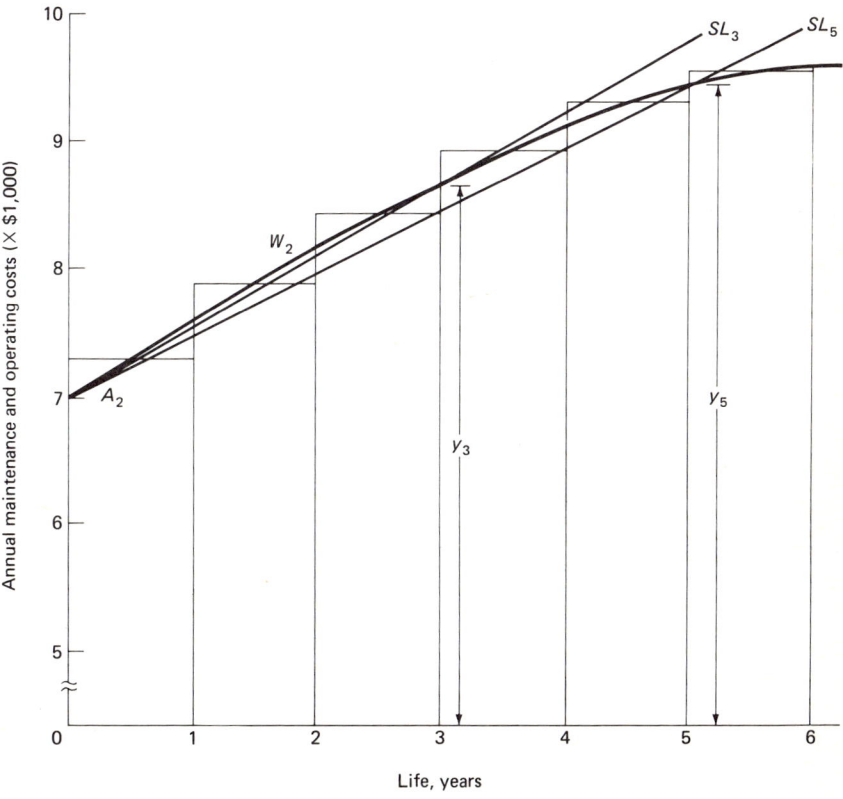

Fig. 5-1 Curve of maintenance and operating costs.

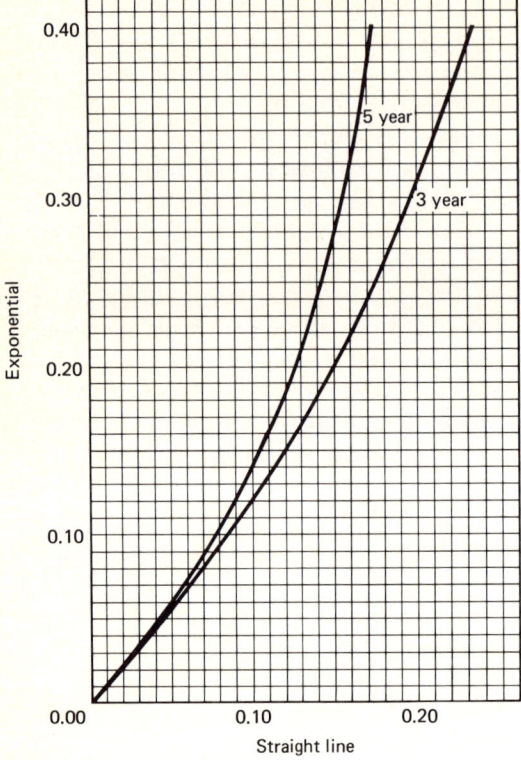

Fig. 5-2 Straight-line equivalents for 3 and 5 years.

rises with age. It is usually convenient to choose B_2 equal to A_2, and this should normally be done.

There are several ways to determine the value of W_2. One method uses a slide rule and templates.[1] Offered herewith is another method of estimating the exponent by referring its straight-line value to Fig. 5-2 or Table 5-5. The table shows the values of these functions for 3 and 5 years computed in increments of 2 percent of the exponent. The straight-line equivalents are obtained by solving the equations

$$SL_x = \frac{1 - e^{-ax}}{x}$$

[1] *Stanford Univ. Constr. Inst. Tech. Rep. 61.*

SOLUTION FOR EQUIPMENT LIFE

Table 5-5 Straight-line equivalents for 3 and 5 years

	Straight line	
Exponent	3 years	5 years
0.00	0.000	0.000
0.02	0.019	0.019
0.04	0.038	0.036
0.06	0.055	0.053
0.08	0.071	0.066
0.10	0.086	0.079
0.12	0.101	0.090
0.14	0.114	0.101
0.16	0.127	0.110
0.18	0.139	0.119
0.20	0.151	0.126

where SL_x = straight-line equivalent for x years
a = exponent (W_2 for M & O curve)
x = number of years

If the life of the machine is estimated to be less than 6 years, the use of SL_3 is suggested. For lives of 6 to 10 years, SL_5 should be used. As an example, refer to Fig. 5-1. Suppose this curve of M & O costs has been sketched. The value of A_2 has been determined to be \$7,000, and the value of y_3 (y intercept at the end of the third year) is \$8,800. The rate for SL_3 is determined by the following procedure:

$$SL_3 = \frac{y_3 - A_2}{3A_2} = \frac{8,800 - 7,000}{3 \times 7,000} = 0.0858$$

When this value is looked up in Table 5-5 or in Fig. 5-2, the value of W_2 is found to be 0.10.

The same kind of procedure can be used for declining curves with a slight modification. To determine S_2, for example, find the fractional loss for 3 (or 5) years, divide by 3 (or 5) to find the SL_x, and look up the answer in the table or figure. Figure 5-3 shows a typical revenue curve which has been sketched from a plot of annual revenues. P_2 has been determined to be \$13,700, and y_3 is \$11,500;

$$SL_3 = \frac{P_2 - y_3}{3P_2} = \frac{13,700 - 11,500}{3 \times 13,700} = 0.0535$$

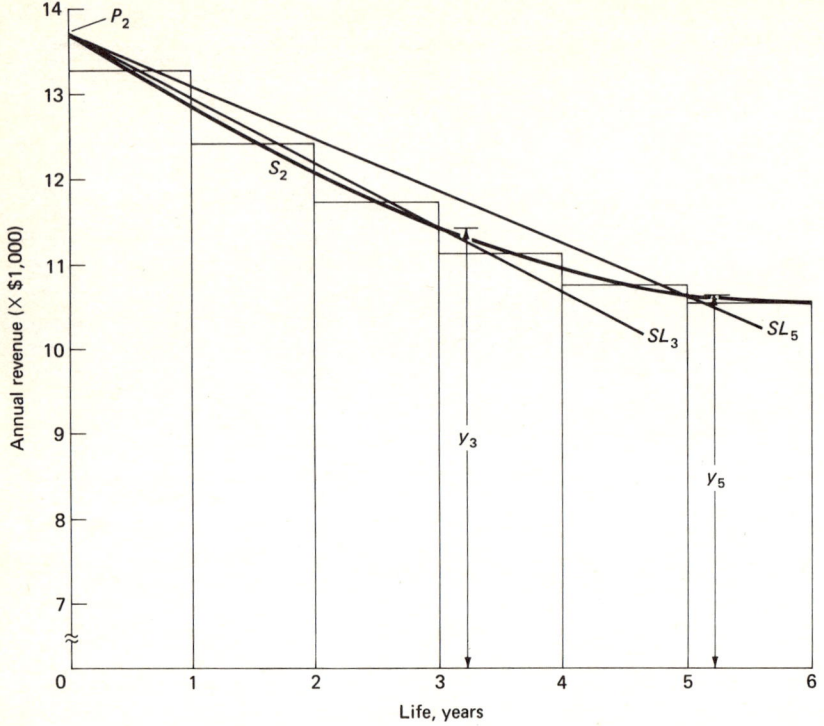

Fig. 5-3 Revenue curve.

Table 5-5 shows the equivalent of $SL_3 = 0.0535$ to be 0.06, which is the value of S_2.

The value of G in the revenue curves should be selected by a consideration of the state of development of the machine at the time of analysis. G will usually be in the range of 2 to 8 percent, which is the range of values in the tables of Appendix 3. G will be at the low end of the range for a machine which has been in development for many years. New types of machines which are in an early stage of development will be toward the high end of the range.

The capital-cost curve can be analyzed in the same fashion as the revenue curve above. Where data are lacking on purchase price and salvage values, guidance can be obtained from the *Green Guide* for construction equipment, *Red Guide* for off-highway trucks and trailers, and the *Blue Book of Rental Rates*[1] or any other good source of market prices. C_4 and K can be ob-

---

[1] These guides are published by the Equipment Guide Book Co., 3980 Fabian Way, Palo Alto, Calif.

SOLUTION FOR EQUIPMENT LIFE

tained by sketching a capital-cost curve from your data. V can be obtained from your records or from data contained in the Factory List Price Indexes published by the Bureau of Labor Statistics, Department of Commerce.

At first, it will seem hard to reason out acceptable values of the variables required for an adequate economic analysis. And yet, one knows intuitively that it must be done in order to get a rational answer. To ignore such things as obsolescence is to deny their presence. To leave them out of the analysis is to equate them to zero. A proper choice always becomes easier with practice and experience. Some genuinely hard brainwork in the beginning will yield dividends over the long haul.

VALUE OF VARIABLES USED IN PROBLEM

It is difficult in an abstract problem to select variables, and no doubt some of the choices will be the subject of argument. Many of the factors which are well known in real life have considerable variance because of geographic location, job conditions, management aptitude, and a host of other controls. The numbers selected for this illustrative problem are felt to be reasonable and should be acceptable for the purpose of demonstrating the use of the tables in finding economic life and time of replacement.

In the solution of the problem, straight-line depreciation will be used for the present machine and double declining balance for future machines. C_7 has been made equal to C_1, but it does not have to be in this case because G_1 will be equal to zero. Remember that when G_1 is required, C_7 must be assumed to be equal to C_1.

Assume that the values of variables have been determined by an analysis of costs as described in the previous section and other source material. These values are as follows:

$A_0^* = 3$	$A_1 = 8{,}000$	$A_2 = 7{,}000$
$B_1 = 8{,}000$	$B_2 = 7{,}000$	$C_1 = 8{,}000$
$C_2 = 17{,}000$	$C_3 = 0.15$	$C_4 = 0.15$
$C_5 = 1{,}500$	$C_6^* = 0.01$	$C_7 = C_1 = 8{,}000$
$F = 0.75$	$G = 0.06$	$I_0^* = 0.02$
$K = 0.70$	$P_1 = 10{,}000$	$P_2 = 13{,}700$
$Q = 13{,}700$	$R_4^* = 0.10$	$R_5 = 0.50$
$R_6^* = 0.0953$	$S_1 = 0.08$	$S_2 = 0.06$
$T_1^* = 1$	$T_2^* = 4$	$T_3 = 3$
$T_4 = 1$	$U = 0.98$	$V = 0.03$
$W_1 = 0.12$	$W_2 = 0.10$	$Z^* = 0.01$

*Table constants.

SOLUTION OF THE PROBLEM

The general procedure in solving a problem of this sort is first to set $N = 0$ and solve for the economic life L by trial and error. The best way to start is to find the profits for $L = 3$, 4, and 5. These will show whether the curve is rising (profits for 3 or 4 are less than 5) or falling (profits for 3 and 4 are greater than 5). If the curve is rising, try $L = 6$ next. If the curve is falling, $L = 3$ will be the maximum since accelerated depreciation and investment credit become available at that time. This should generally locate the maximum profit in 3, or possibly 4, trials. Once the year of maximum profit is found, L is set to this value and the profit is found with $N = 1$. If this profit is less than that found with $N = 0$ above, the time to replace is now and you need go no

Figure 5-4 Example problem: N = 0, L = 3, 4, 5, and 6

			$L = 3$	
Factor	Base variable	Table variable	Table value	Equation value
r_2*,†	$P_2 + Q = 27{,}400$	$S_2 = 0.06$	9.64313	264,222
r_3*,†	$-Q = -13{,}700$	$G = 0.06$	6.43873	−88,211
		$S_2 = 0.06$		
e_3‡,§	$A_2 = 7{,}000$		9.63205	67,424
e_4‡,§	$B_2 = 7{,}000$		8.90920	62,364
e_5‡,§	$-B_2 = -7{,}000$	$W_2 = 0.10$	7.75194	−54,264
e_7§	$C_2 = 17{,}000$	$K = 0.70$	3.73465	63,489
		$V = 0.03$		
		$C_4 = 0.15$		
e_{11}‡,§	$C_5 = 1{,}500$	$T_3 = 3$	0.00000	0
d_2‡ (d_6)	$C_2 = 17{,}000$	$V = 0.03$	4.95304	84,202
i_2‡,§	$0.75 C_2 = 12{,}750$	$V = 0.03$	1.42534	18,173
g_2*	$C_2 = 17{,}000$	$V = 0.03$	1.73825	29,550
		$K = 0.70$		
		$C_4 = 0.15$		
R'*				205,561
E'‡				177,899
ic_{11}	PIC × C_2	$V = 0.03$	1.00000	397
ic_{12}	−PIC × C_1	$C_3 = 0.15$	1.00000	−187
ic_2	PIC × C_2	$K = 0.70$	2.73466	1,086
		$V = 0.03$		
		$C_4 = 0.15$		
IC	$ic_{11} + ic_{12} + ic_2$			1,296
e_9§	$R_5(R' - E') - $ IC			12,535
R_0†				176,011
E_0§				169,721
PW_L	$R_0 - E_0$			6,290

*Add these to get R'.   †Add these to get R_0.
‡Add these to get E'.   §Add these to get E_0.

SOLUTION FOR EQUIPMENT LIFE

further unless you are curious to see how much more you lose by retaining the present machine 2 years more. If the profit curve is rising (profit for $N = 1$ is greater than $N = 0$), then solve for profit with $N = 2$. If the profit for $N = 2$ is less than for $N = 1$ and the profit for $N = 1$ is greater than the profit for $N = 0$, then the time to replace is in 1 year more (that is, $N = 1$). If the profit for $N = 2$ is greater than for $N = 1$, an analysis should be made annually until the time to replace appears.

ECONOMIC LIFE

The format for solution of the economic life of future machines is shown in Fig. 5-4. This analysis is made with $N = 0$, which means replace now. With $N = 0$ all equations for the present machine

$L = 4$		$L = 5$		$L = 6$	
Table value	Equation value	Table value	Equation value	Table value	Equation value
9.39898	257,532	9.17220	251,318	8.96165	245,549
6.43874	−88,211	6.43873	−88,211	6.43874	−88,211
9.67469	67,723	9.71592	68,011	9.75574	68,290
8.98853	62,920	9.06540	63,458	9.13978	63,978
7.50225	−52,516	7.27500	−50,925	7.06834	−49,478
3.20742	54,526	2.85160	48,477	2.58809	43,998
2.18553	3,278	1.83534	2,753	1.60401	2,406
3.58230	60,899	2.80050	47,609	2.29598	39,031
1.39380	17,771	1.15021	14,665	0.98840	12,602
0.97083	16,504	0.58106	9,878	0.36084	6,134
	185,825		172,985		163,472
	160,075		145,571		136,829
1.00000	397	1.00000	793	1.00000	793
1.00000	−187	1.00000	−373	1.00000	−373
2.20743	876	1.85160	1,468	1.58809	1,260
	1,086		1,888		1,680
	11,789		11,819		11,642
	169,321		163,107		157,338
	165,491		158,258		153,438
	3,830		4,849		3,900

$$\text{PIC} = \frac{0.07n}{3} \quad \text{where} \quad L < 3, n = 0$$
$$3 \leq L < 5, n = 1$$
$$5 \leq L < 7, n = 2$$
$$L \geq 7, n = 3$$

become equal to 0 and hence may be omitted from the solution. The format for solving economic life does not require r_1, e_1, e_2, e_6, e_8, e_{10}, d_1, d_5, g_1, or i_1. The first step is to record the values of all base and table variables. Next, the tables of Appendix 3 are used to find and record on the form all table values for the factors required for solutions when $L = 3$, 4, and 5. Interpolate where necessary to obtain values to the nearest percentile. Multiply base variables by table values to obtain equation values and fill in that column down through the investment-credit row. Sum to obtain R' and E'. Next figure E_9, the income tax due, by combining R', E', and R_5 and subtracting IC. Remember in computing PIC that under the present law one-third of 7 percent will be used for $L = 3$ and 4 and two-thirds of 7 percent will be used for $L = 5$. In other words, $\text{PIC}_{3,4} = 0.0233$ and $\text{PIC}_5 = 0.0467$. Combine equations now to obtain R_0, E_0, and then PW_L (present worth of profits at year L).

All the operations above having been performed for $L = 3$, 4, and 5, it is observed that $\text{PW}_3 = \$6{,}290$, $\text{PW}_4 = \$3{,}830$, and $\text{PW}_5 = \$4{,}849$. The profit curve is rising, and so the next trial will be made with $L = 6$. Again the procedure outlined above is executed, and $\text{PW}_6 = \$3{,}900$. Note that the second overhaul due at the end of the sixth year has not been performed, but if equipment is retained to the seventh year, it must be overhauled. It is now clear that the maximum profits will be attained by replacement of future machines at 3-year intervals; hence, the economic life of these machines is 3 years. It is interesting to note that an analysis of these machines under the old rules where investment credit was acquired at 4, 6, and 8 years yielded an economic life of 4.1 years. There was just the proper combination of costs to make the new rules of 3, 5, and 7 years shorten the economic life by a year—just the effect intended by the law.

REPLACEMENT LIFE

The format for solution of replacement life is shown in Fig. 5-5. It will be noted that all the equations are required for this solution but that no more than one or two trials will be required. Proceed in the same order as in determining economic life except that now $L = 3$. Solve the equations first with $N = 1$ (the solution for $N = 0$ has already been accomplished in Fig. 5-4). Record all base variables and table variables. Next look up the set of table values for all factors. Now multiply base variables by table values to obtain equation values. Sum to obtain R' and E'; com-

Fig. 5-5 Example problem: $L = 3$, $N = 1$.

			$L = 3, N = 1$	
Factor	Base variable	Table variables	Table value	Equation value
r_1*,†	$P_1 = 10,000$	$S_1 = 0.08$	0.91725	9,173
r_2*,†	$P_2 + Q = 27,400$	$S_2 = 0.06$	8.76648	240,201
r_3*,†	$-Q = -13,700$	$G = 0.06$	5.51252	−75,521
		$S_2 = 0.06$		
e_1‡,§	$A_1 + B_1 = 16,000$		0.95382	15,261
e_2‡,§	$-B_1 = -8,000$	$W_1 = 0.12$	0.89967	−7,197
e_3‡,§	$A_2 = 7,000$		8.66928	60,678
e_4‡,§	$B_2 = 7,000$		8.09927	56,695
e_5‡,§	$-B_2 = -7,000$	$W_2 = 0.10$	7.04722	−49,331
e_6§	$C_1 = 8,000$	$C_3 = 0.15$	0.21750	1,740
e_7§	$C_2 = 17,000$	$K = 0.70$	3.49854	59,475
		$V = 0.03$		
		$C_4 = 0.15$		
e_{10}‡,§	$C_5 = 1,500$	$T_3 = 3$	0.00000	0
		$T_4 = 1$		
e_{11}‡,§	$C_5 = 1,500$	$T_3 = 3$	0.00000	0
d_1‡(d_5)	$C_7 = 8,000$	$C_3 = 0.15$	0.11510	892
d_2‡(d_6)	$C_2 = 17,000$	$K = 0.70$	4.63989	78,878
		$V = 0.03$		
		$C_4 = 0.15$		
i_1‡,§				
i_1‡,§	$0.75 C_2 = 12,750$		0.09445	1,204
i_2‡,§	$0.75 C_2 = 12,750$	$V = 0.03$	1.33522	17,024
g_1*	$C_1 = 8,000$	$C_3 = 0.15$	0.00000	0
g_2*	$C_2 = 17,000$	$V = 0.03$	1.62835	27,682
		$K = 0.70$		
		$C_4 = 0.15$		
R'*				201,535
E'‡				174,104
ic_{11}	PIC $\times C_2$	$V = 0.03$	0.93678	372
ic_{12}	$-$PIC $\times C_1$	$C_3 = 0.15$	0.78251	−146
ic_2	PIC $\times C_2$	$K = 0.70$	2.56177	1,017
		$V = 0.03$		
		$C_4 = 0.15$		
IC	$ic_{11} + ic_{12} + ic_2$			1,243
e_8, e_9 §	$R_5(R' - E') - $ IC			12,473
R_0†				173,853
E_0§				168,022
PW_N	$R_0 - E_0$			5,831

*Add these to get R'.
†Add these to get R_0.
‡Add these to get E'.
§Add these to get E_0.

$$\text{PIC} = \frac{0.07n}{3} \quad \text{where} \quad \begin{array}{l} L < 3, n = 0 \\ 3 \leq L < 5, n = 1 \\ 5 \leq L < 7, n = 2 \\ l \geq 7, n = 3 \end{array}$$

Table 5-6 Present worth of profits

L	N 0	N 1
3	$6,290	$5,831
4	3,830	
5	4,849	
6	3,900	

bine these with R_5 and IC to obtain E_8 and E_9. Finally the proper equations are combined to find R_0, E_0, and PW_N.

After these operations are completed for $L = 3$ and $N = 1$, it will be observed that $PW_1 = \$5,831$. That is somewhat less than $PW_0 = \$6,290$. With these results, we can conclude that the time to replace is now and positively before the end of another year.

SUMMARY OF SOLUTION

The solution of this problem by the tables in Appendix 3 has produced the results in Table 5-6. From these results, the two

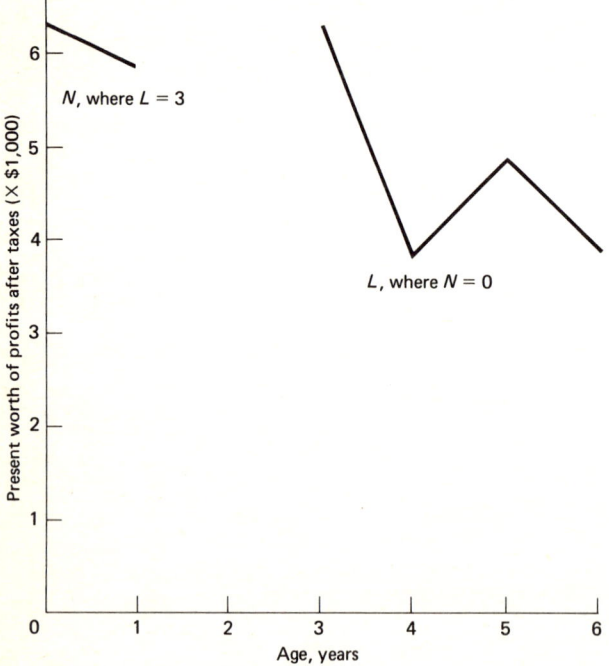

Fig. 5-6 Values of L and N.

SOLUTION FOR EQUIPMENT LIFE

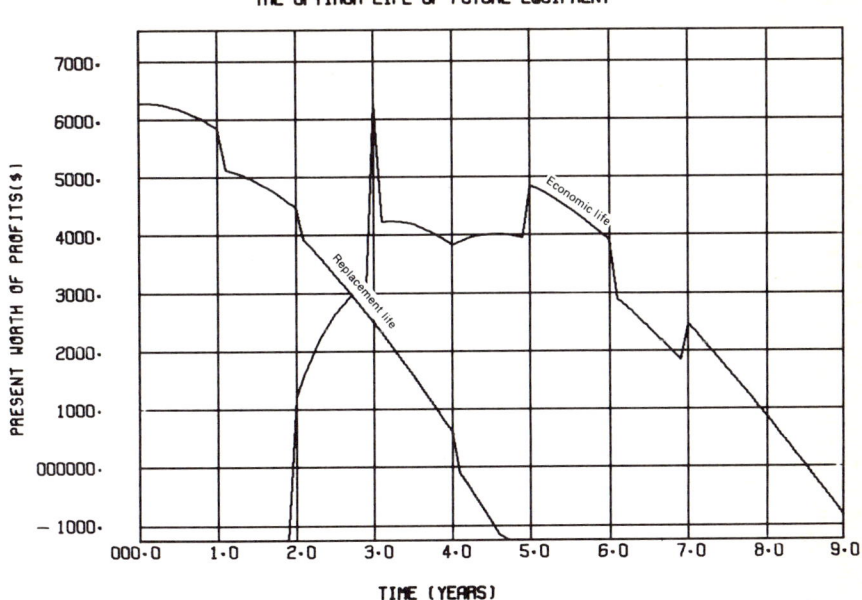

Fig. 5-7 Example problem: computer plot.

curves in Fig. 5-6 can be drawn. It is clear that the economic life of future machines is about 3 years and the time to replace the present machine is now.

A computer run with the same input data verifies these decisions and sheds further light on the shapes of the two curves. Figure 5-7 is the plot of the computer solution to the example problem. The computer has optimized at $N = 0.0$ and $L = 3.0$, which is more precise than can be done with the tables. The maximum profit at these lives is $6,281.08. The plot shows that the maximum at 3.0 years survives only until the overhaul is performed. If the overhaul is in fact accomplished, the next high profit occurs at 5.0 years, when the second increment of investment credit is attained. Practically, the results are compatible, and decisions would be the same with either solution. Figure 5-8 is the computer output for $N = 1$ and $L = 3$. This verifies all of the equation values of the tabular solution in Fig. 5-5. The present worth of profits with investment credit (PTI) in Fig. 5-8 in the amount of $5,834.44 confirms the result of $5,831 obtained in the tabular solution.

Figure 5-9 is a partial list of the values of the present worth of profits after tax for use in plotting the curves shown in Fig. 5-7. These are values for $N = 0.0$ and $L = 3.0$ for use in the op-

```
       THE GIVEN DATA           THE TEST RESULTS
                               FOR N= 1.0 AND L= 3.0
-------------------------------------------------------

           A1=   8000.000         D1  =       0.000
           A2=   7000.000         D2  =   78878.120
           B1=   8000.000         D3  =       0.000
           B2=   7000.000         D4  =       1.099
           C1=   8000.000         D5  =     921.110
           C2=  17000.000         D6  =       0.000
           C3=      0.150         G1  =       0.000
           C4=      0.150         G2  =   27668.420
           C5=   1500.000         E1  =   15261.170
           C6=      0.010         E2  =   -7197.379
           C7=   8000.000         E3  =   60684.950
           D3=      0.000         E4  =   56694.900
           D4=      1.099         E5  =  -49330.58
           F =      0.750         F6  =    1740.304
           G =      0.060         E7  =   59488.610
           T0=      0.020         E8  =    -508.322
           L =      3.000         E9  =   14198.590
           N =      1.000         E10 =       0.000
           P1=  10000.000         E11 =       0.000
           P2=  13700.000         TC  =    1242.008
                                  R1  =    9172.500
           R4=      0.100         R2  =  240201.600
           R5=      0.500         R3  =  -75521.50
           P6=      0.095
           S1=      0.080
           S2=      0.060
           T1=      1.000         I1  =    1204.236
           T2=      4.000         I2  =   17024.070
           T3=      3.000         E0  =  169260.200
           T4=      1.000         R0  =  173852.600
           U =      0.980
           V =      0.030         PT  =    4592.438
           W1=      0.120         PTI =    5834.441
           K =      0.700         TC1 =     225.529
           W2=      0.100         TC2 =    1016.479
           Z =      0.010
           C =  13700.000
```

Fig. 5-8 Example problem: computer output for test run.

timal solution. The ordinates of the economic-life curve appear in the column labeled "Future" and the ordinates for the replacement life curve are in the column labeled "Present."

Altogether one may conclude that the tabular results are satisfactory in this case for determination of economic life and replacement life. It should be noted that usually the maximum value on the curves is rather flat for at least a year *unless* some event occurs which causes it to jump up or drop. It often happens that these events are the signal to retire the machine. For example, major overhauls cause a drop in profits, so that the time to dispose of the machine is prior to a major overhaul, preferably when the salvage value can be maximized. Investment credit will cause a jump in profits; wait until the third, fifth, or seventh year if you are close to it. Once the approximate time to replace is determined, the deed becomes a matter of opportunity, which

SOLUTION FOR EQUIPMENT LIFE

may take place over a period of several months, seizing the proper time to maximize salvage value, use double-declining-balance depreciation (at least a 3-year life), obtain investment credit (3, 5, and 7 years), avoid overhaul, or make a good deal on trade-in. This final decision can be made more comfortably and confidently if you have a good economic analysis to back up your decision.

LIFE	FUTURE	PRESENT
	N= 0.0	L= 3.0
0.0	-17000.00	6281.08
0.1	-370464.90	6278.14
0.2	-175522.20	6265.04
0.3	-110727.90	6242.39
0.4	-78470.69	6210.25
0.5	-59229.06	6169.09
0.6	-46494.69	6119.00
0.7	-37478.63	6060.13
0.8	-30786.69	5992.91
0.9	-25644.25	5917.60
1.0	-21536.38	5834.45
1.1	-18317.13	5118.01
1.2	-15639.25	5075.24
1.3	-13416.38	5024.75
1.4	-11550.81	4966.43
1.5	-9971.13	4900.77
1.6	-8623.56	4827.72
1.7	-7467.00	4747.77
1.8	-6469.81	4660.81
1.9	-5606.44	4567.20
2.0	1210.31	4467.15
2.1	1586.88	3923.58
2.2	1915.50	3771.81
2.3	2198.63	3618.53
2.4	2439.88	3463.30
2.5	2642.81	3305.75
2.6	2810.63	3145.76
2.7	2946.88	2983.01
2.8	3054.50	2817.38
2.9	3135.88	2648.87
3.0	6281.08	2477.36
3.1	4228.28	2302.79
3.2	4238.04	2125.28
3.3	4231.67	1944.82
3.4	4210.09	1761.31
3.5	4174.52	1574.99
3.6	4126.13	1385.61
3.7	4065.84	1193.41
3.8	3994.60	998.60
3.9	3913.22	800.98
4.0	3822.47	600.93
4.1	3893.24	-114.12
4.2	3947.04	-319.34
4.3	3985.28	-526.87
4.4	4008.80	-736.83
4.5	4018.94	-948.97
4.6	4016.48	-1163.41
4.7	4002.34	-1379.65
4.8	3977.37	-1598.00
4.9	3942.29	-1818.35
5.0	4842.54	-2040.55
5.1	4777.46	-2264.50
5.2	4704.48	-2490.05
5.3	4624.21	-2717.22
5.4	4537.11	-2946.19

Fig. 5-9 Example problem: computer output for plot (partial).

6
Cost, Time, and Production Records

IMPORTANCE OF RECORDS

Cost, time, and production records are absolutely essential for the successful operation of any business. These records are frequently neglected in equipment operations because the owner feels they are too costly, too time-consuming, and simply too much trouble to keep. *Business Management* reported in November 1964 that fewer than 15 percent of the private carriers in the United States have any idea what their trucking costs are. In a study of truck costs made by the University Research Center in 1965,[1] it is stated that an examination of trucking costs among 1,338 companies indicates that only approximately 5 percent of them keep sufficient records to reveal accurate costs. While these statistics are for cross-country truckers, it is most

---

[1] *"Truck Costs,"* p. 6, University Research Center, Inc., Chicago, 1965.

COST, TIME, AND PRODUCTION RECORDS

likely that contractors have no better batting average in the realm of equipment cost keeping.

WHY RECORDS ARE KEPT

These cost, time, and production records need to be kept for several reasons:

Cost control of the job in progress   They tell you where your costs are now, where they have been in the past, and where they should be at some future time.

Financial condition   Overruns and underruns show where you erred in your past estimates of cost. This helps to spotlight trouble so that corrective measures can be taken as soon as possible.

Progress on the job   How much has been spent to date and how much has been accomplished will help to keep the work on schedule.

Standards of cost, time, and production   These are fundamental to the successful continuation of the enterprise. Learning is a constant process. These records document past performance under stated conditions and are useful in estimating future costs and schedules.

Basis for litigation   Claims for extra costs must be supported by documentary evidence of costs and work performed. They may also be useful in protecting the owner against damage suits brought as a result of his operations.

Establish tax liability   Deductible expenses, both operating and ownership, must be documented for the tax records. Gain on sale on the individual machine, for example, is subject to recapture at the full tax rate and must be reported as income.

Data for economic analysis   Without these data it is impossible to optimize economic life and replacement timing. Decisions made on a sound basis of costs and performance will be on a higher level of confidence than those made intuitively.

HOW COSTS RECORDS ARE USED

Cost records are used directly in keeping the costs of the enterprise. There are other uses, however, in which these costs are transformed or manipulated before they become useful for certain purposes. These cost forms are not always consistent or compatible with the accounting records.

One of the stated purposes of engineering economics is to reduce everything to a common denominator, the dollar. Sometimes it is necessary to impute costs for intangibles, since these will not be found in the books of accounts. Such factors as obsolescence, downtime, breakdown, and loss of goodwill resulting from poor operation all may contribute to costs required for decision making. In minimizing costs to obtain economic life, it is necessary to include these intangibles among the other costs of equipment. To ignore them is to deny them.

Another example of deviation from the time and cost records is the use of straight-line depreciation for figuring capital cost in the determination of a use or rental rate. While it is recognized that accelerated depreciation has certain advantages and should be used for tax purposes, the early write-off would discourage the use of new equipment if it were used to compute a use rate. Therefore, once the economic life is determined by using an accelerated method, this life and a reasonable salvage value are used to determine the straight-line depreciation write-off for calculating the use rate.

Maintenance and operating costs, too, are averaged out over the life of the machine to determine the use rate. Otherwise, no one would want to use the older machines, which inevitably have higher maintenance and operating costs. For this reason, the total costs over the economic life are averaged out to determine the hourly charge for use or rental.

In the economic analysis of a machine, it is very important to use the costs at the rate of accrual, since money has a time value. The cost records are then used directly for this purpose and indirectly for the calculation of use or rental rates.

WHAT RECORDS ARE KEPT

Records of cost, time, and production must be complete if they are to be used properly. That means that all costs, time, and production must be kept. Of fundamental importance for economic analysis is a record of these factors by years. This may be by calendar year, fiscal year, or year of age, provided the records are not distorted by unusual tax write-offs or other fiscal operators. In the latter case, these factors will have to be recognized at the time of analysis. The detail with which the records are kept will depend on the purposes served. It is a waste of time and money to keep records in infinite detail unless they serve a useful purpose in that form. To that end, the alert manager should seek to determine why each statistic is collected and how it will be used. While it may be sufficient to get a total of maintenance costs annually for the determination of economic life, it may be necessary to get a monthly report of high maintenance cost to see whether equipment is being abused or operated beyond its capacity. A monthly printout of high fuel costs may reveal fuel thefts or poor engine operation. By all means, keep all costs but only in the detail necessary to implement management control and fiscal responsibility.

THE NATURE OF EQUIPMENT COSTS

Equipment costs can be classified as fixed, variable, and mixed. Fixed costs are related to calendar time. They accrue with the passage of time and are independent of the amount of operation of the machine. Variable costs are related to hours of operation and are limited by calendar time only because there are 24 hours in a day and 365 days in a year. Mixed costs are related to both calendar time and operating hours. Because of the inherent nature of these costs—some accruing with calendar time, some with operating hours, and some with both—the combination of all three to obtain cost per hour is complex and highly subjective. In fact, any statement of cost per hour should be followed by the conditions of calendar time and operating hours under which it was computed, e.g., average annual cost per hour based on 1,500 hours of operation.

Fixed costs, being time-related, usually include those which are common to ownership, often called ownership costs. All fixed costs are ownership costs, but the converse is not true. Some ownership costs are mixed, and there is difficulty in combining them with the fixed costs of ownership. The true fixed costs are those like insurance, obsolescence, property taxes, licenses, fees, and interest charges. They accrue whether the equipment is in use or not. Customarily charges for these costs are made on the basis of a 30-day month and spread over the number of months of utilization of the equipment.

Variable costs will increase as the machine ages, but they occur only when the machine is being operated. These costs usually occur at short intervals, such as the time between refueling or the time interval of periodic maintenance. These intervals are usually short, a period of several hours to several weeks, and should be treated as hourly operating costs for analytical purposes. One set of costs, however, is related to operating hours, although the costs may occur years apart. These are the costs of overhaul and major repairs. They should be charged for by hours of operation and accrued for use at the time of overhaul or major repair. In analysis they should be properly timed to their occurrence and not treated as an hourly charge.

Although they are variable costs, they are usually considered as an ownership cost because of the interval. Fuel, tires, operating labor, field repairs, and lubrication costs are commonly treated as variable costs and charged by hour of operation.

Mixed costs are the most complex and the most difficult to handle. Since they are mixed, it is not clear whether they should be charged off as fixed costs or variable costs. When either type is dominant, the choice is simple, but this condition does not always obtain. Depreciation, for example, consists of two parts, deterioration and obsolescence. Deterioration is physical and due both to wear and tear and to aging. It usually occurs with hours of use but may also occur as a result of outside storage without proper preservation. Here the assignment of the machine will determine whether its physical decline is due to wear and tear or to aging, the former related to hours of operation and the latter to storage conditions and calendar time. Obsolescence, on the other hand, is related generally to calendar time. Use is not a factor. It accrues with the upward creep of technology, sparked from time to time by some innovative breakthrough. It may be treated as a fixed cost.

In depreciation we are asked to combine these two divergent factors. Their combination is not simple. If a machine is operated 3,000 hours/year, it certainly depreciates faster than one of the same vintage which is operated only 500 hours/year. Depreciation is a mathematical method of treating capital runoff which really is measured by the decreasing market value of the machine as it is used up in service. It is best to treat depreciation as a fixed cost, always keeping in mind that it should be adjusted according to the hours of operation.

DIFFICULTY IN COMBINING COSTS

It is the differing structure of the three kinds of costs discussed above that makes it difficult to combine them. Since the costs are based on two different rates of accrual, calendar time and operating hours, it is necessary to assume the number of hours of operation in future years in order to arrive at an estimated total direct cost per hour for a machine. The combination of over a dozen different costs to arrive at this hourly cost suggests a trial-and-error solution of various operating schedules in order to minimize the hourly cost. Studies by the author indicate that two 10-hour shifts per day are generally more economical than one 8-hour shift. Of course, the quality and availability of labor and many other local factors make a trial solution necessary for each set of job conditions.

As the balance of labor and equipment costs changes, new solutions to the problem of minimizing hourly cost must be sought. The cost of new, larger, and more productive machines has weighted cost reduction toward the increased utilization of the machine. This is particularly true in areas where there are few good earth-moving days per year. Contractors in the Pacific Northwest usually figure about 100 earth-moving days per year because of adverse weather conditions. The use of two 10-hour shifts per day in this area results in about 2,000 scheduled hours per year and much lower cost than would result from single-shift operation.

USE OF HISTORICAL COSTS

All the decisions which must be made in a business are concerned with the future. Hence, the most important use of costs relates to things that will happen in the future. This is true in regard to equipment whether you are trying to set up a rental or use rate, predict the economic life of a machine, or determine the time to replace a present machine. Historical costs are principally used for predicting future costs. A secondary use is to compute your tax liability.

Human judgment is required to interpret these historical costs and apply them to future situations. The computer can spew out all kinds of statistical information which may supply some guidance, but the man making the decisions is the most important adjunct to good policy. Where the future job is similar to a past job, statistical data from the latter are extremely valuable. For this reason, good historical costs are important.

The manager must always be wary of sunk costs. Entirely too many decisions are made on an emotional basis of likes and dislikes. Past performance and costs are important only as they affect the future. High expenditures on the maintenance of a machine are sunk costs. If it were wise to dispose of the machine, the time to do it would be before the high maintenance cost occurred. Cost trends are helpful, however, and gradually increasing maintenance costs should lead the manager to some sound analysis to determine whether the machine should be replaced or retained. The method of analysis outlined in previous chapters of this book will be helpful in this regard.

It is important in respect to the use of historical costs that they be in sufficient detail to be adaptable to new situations. For example, fuel consumption should be related to severity of the type of work. Ripping rock with a crawler tractor always uses more fuel than pulling sheepsfoot rollers. Tire and track costs, too, must be related to job conditions. The life of tires or tracks may vary from 500 hours up to 5,000 hours, a tenfold difference in cost. In spite of the caution which must be used in handling historical costs, there is no substitute for them. Perhaps standard costs, which would necessarily derive from many historical costs, will someday be available to assist those who need them. Until that day arrives, the contractor's own records will have to provide the basis for estimating future costs.

USE OF COST, TIME, AND PRODUCTION ACCOUNTS

Cost, time, and production accounts are of no avail unless they are properly utilized. The cooperation and understanding of all members of the organization are required to operate a successful record program. Information must be recorded accurately and in a timely fashion without bias. The time to ferret out errors and distortion in reporting is before the information is fed into the system. Once the data are fed into a bank, it is unlikely that errors will be found except by accident. To that end, major expenditures should be checked by an equipment-cost engineer before being entered into the data bank. Preferably there should be a check on all data submitted by the next line of supervision before they are entered in the system. If there are fewer than 500 pieces of equipment in the plant account, a manual system may be adequate. Any manual system which can be operated by three persons or fewer can be implemented more economically than a computerized system. When more than three employees are required full time, the advantages of computer processing become more apparent.

WHAT EQUIPMENT SHOULD BE INCLUDED

All equipment should be in the plant account and records kept on it. All machines with a value of more than $500 should have individual accounts. Less expensive equipment such as small air compressors, generators, pumps, and other auxiliaries should be maintained in vintage group accounts of like makes and models. In this way, cost and production information will be available to aid in the selection of replacement machines. Equipment should be identified in the accounts by the contractor's number on the machine. The development of a numbering system for inventory will be discussed in a subsequent chapter.

COLLECTION OF DATA

Any system that is adequate may be used for data collection, but keep it simple. All information which can be printed on the form

COST, TIME, AND PRODUCTION RECORDS

to be filled out by the employee should be included in the form design. Accounting systems and data-collection forms have been described by Brock, Hackney, Stewart, the ASCE cost control manual, and numerous other publications.[1] Any new system which is adopted by the contractor will take at least 1 year to implement and probably 2 years to make effective operationally. The cost of operating a computerized equipment-management information system is a matter of concern to those contractors who have always done the work manually. It has been stated that present costs for routine processing of equipment information in one company are about 0.2 percent of equipment expenditures.[2]

Data collection should first be concentrated on the equipment of highest cost. Another interesting observation is that approximately one-fourth of the equipment owned by one company accounts for three-fourths of the total equipment cost.[3] Once data collection and processing are successfully implemented for the high-cost equipment, it should be extended by groups over the entire equipment spectrum.

CHART OF ACCOUNTS

The listing of a chart of accounts is common for cost coding. Not so common is a chart of accounts for recording time and performance. Appendix 4 is a chart of accounts of all three factors. The accounts are numbered to be adaptable to either manual or computer storage.

Cost, time, and production items should be coded according to the following key:

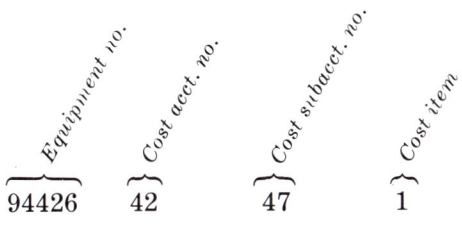

94426	Contractor's Equipment Number
.42	Field repair
47	Cooling system
1	Labor

---

[1]See list of references at end of chapter.
[2]Elmer A. Cox, Information Needs for Controlling Equipment Costs, *Highw. Res. Board Highw. Res. Rec.* 278, p. 46, Washington, August 1969.
[3]Ibid., p. 46.

The suggested chart of accounts in Appendix 4 may be expanded or contracted further to accommodate the requirements of the company. A listing of the basic cost accounts is the minimum breakdown of information that should be kept on any machine. Table 6-1 is a listing of these accounts. Accounts should be kept for each machine individually. Information can be collected by a report system based on daily equipment-operator reports and daily foreman reports.

The key to any control system is the breakdown of information into a useful form. The accounts from .10 to .19 are used to correlate the other data collected and present them in a form useful to management. These figures will be expressed as costs per operating hour for off-highway equipment and as costs per operating hour or cost per mile for highway equipment. Every reasonable effort must be made to record all costs, no matter how small. Costs should be recorded to the nearest cent, time to the nearest $\frac{1}{10}$ hour, and production to the nearest unit (say, bank cubic yard).

Where the cost account number does not give sufficient detail, a cost subaccount number has been used. If this is insuf-

Table 6-1 Chart of basic accounts

Cost	.10	Total profit (loss)
	.20	Total revenue
	.30	Total fixed costs
	.31	Depreciation
	.32	Interest, insurance, and taxes
	.33	Storage and security
	.34	Fees, licenses, and fines
	.35	Moving costs
	.36	Overhead
	.40	Total variable costs
	.41	Field repairs
	.43	Fuel
	.44	Tires or tracks
	.45	Operating labor
	.46	Field supervision
	.47	Major repairs
	.48	Overhauls
Time	.50	Total shift hours
	.51	Idle time
	.52	Total scheduled hours
	.53	Downtime
	.54	Total operating hours
Production	.60	Total units produced

ficient, a two-digit cost sub-subaccount number may be used. The last digit in the cost code is for the type of cost, i.e., labor, material, etc. The last digit of the time account is to segregate the waits for parts and repair.

The purpose of the time accounts is to record the total shift hours, total scheduled hours, total operating hours, and the delays which made them differ. Expressed mathematically, these factors are (numbers are for single-shift operation):

H = total shift hours per year (253 working days × 8 hours per working day = 2,024 hours)
S = total scheduled hours (total number of days work is scheduled × 8 hours per working day)
O = total operating hours (sum of all hours the machine has actually worked)
U = utilization
A = availability
I = idle time, no work scheduled
D = downtime, machine not able to work

Now it is possible to show the algebraic relationship of the above variables.

$$I = H - S$$
$$D = S - O$$
$$\%U = \frac{100S}{H}$$
$$\%A = \frac{100O}{S}$$

The above relationships are often used but seldom defined. With a set of time accounts like those in Appendix 4, it should be possible to standardize these terms. They will follow the definitions above whenever discussed in this text.

CLEARING ACCOUNTS

Depending on the nature of the contracts performed, the contractor may want to treat his equipment accounts as clearing or suspense accounts. The accounts are distributed or cleared to other accounts as the job progresses, so that they are totally cleared or in balance when the job is finished and the books are closed out. Clearing accounts ordinarily should not be used for minor items of equipment such as pumps and generators. It should be remembered that clearing the accounts imposes an additional load on the accounting system: the effort must be justified by the additional return from using them.

USING ACTUAL COST TO OBTAIN COST AT CONSTANT DEMAND

Some enterprises employing equipment are characterized by a fluctuating demand. When hourly costs are reckoned in this kind of business, it is hard to make sense out of either hourly costs or annual costs. Again, we are confronted with the varying mix of fixed and variable costs as the demand changes. A good example is a highway contractor whose work load changes with the ebb and flow of the economy and his competition. Plotted annual costs for maintenance and operation in this situation are not rational, and a method is needed to hypothesize the cost curves under an average demand. Such a method will be offered here.

First, fixed and variable costs must be separated and listed annually from date of purchase out to a life of 5 years. Scheduled hours and operating hours should be tabulated along with these costs to reflect the decreasing availability of the machine as it aged in use. If the scheduled hours and availability have been recorded, the operating hours per year can be determined by the relationships defined previously.

Since availability and operating hours are dependent variables, the annual availability will depend on the actual number of hours worked that year. Figure 6-1 shows curves of availabil-

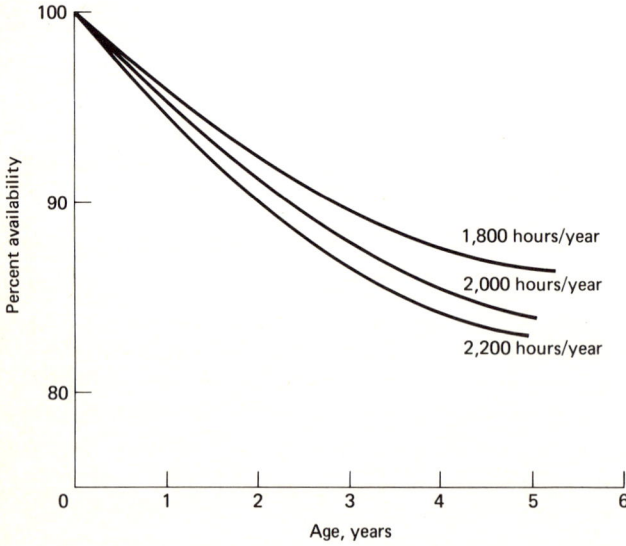

Fig. 6-1 Variation of availability with change in scheduled hours.

ity for a machine scheduled to operate at a rate of 1,800, 2,000, and 2,200 hours/year. Availability will also decline as the number of shifts are increased, for there will be less time outside of working hours for maintenance and repairs. Stewart[1] says that downtime for maintenance should be about 5 to 7 percent when performed during the day shift and 2 to 3 percent when scheduled during the night shift. He further states that downtime for parts should not exceed 1.3 percent of scheduled time.[2] The best device is not to work with availability but to plot total operating hours versus total scheduled hours and work from this graph. This will be illustrated in the example problem which follows.

Once the annual costs and operating hours are known, the cost per operating hour is figured for each year of actual cost. In order to level out the cost, the total number of operating hours is divided by the number of years to obtain an average number of operating hours per year. This is the best estimate of demand unless other facts are known which would cause a variation. If the economic forecasts predict an increase in average demand over the next few years, the average number of operating hours may be increased to reflect this situation.

Now, by using the average number of working hours forecast, the operating-hour curve, and the variable hourly costs, a new maintenance- and operating-cost curve can be synthesized which will reflect the best forecast of a future-cost curve. In order to make this explanation clearer, an example problem will be used to illustrate this technique.

EXAMPLE PROBLEM

The best way to explain the details of handling actual mixed costs with a variable demand is to work a problem related to this situation. A contractor is using a small mobile generator for lighting night operations. His scheduled use has varied over the life of his present machine, and he would like to level out the costs at an average demand so he can determine the parameters of his M & O cost curve. Fortunately, he has collected the tabulated data over the past 5 years. He has scheduled the machine an average of 2,000 hours/year and figures that to be a good estimate of his future schedule. Average availability has been about 90 percent ($^{9,040}/_{10,000}$), and his costs have averaged

---

[1] R. S. Stewart, Motor Vehicle Fleet Management, *Am. Public Works Assoc. Res. Found. Spec. Rep.* 37, Chicago, 1970.
[2] Ibid., p. 9.

Year	Scheduled hours	Operating hours	M & O costs
1	2,210	2,170	$1,430
2	1,850	1,710	1,320
3	2,200	1,930	1,660
4	2,130	1,840	1,800
5	1,610	1,390	1,510
Total	10,000	9,040	$7,720
Average per year	2,000	1,808	$1,544

about 85 cents per operating hour. His fixed costs amount to $300/year, regardless of the amount of operation. What values should he use for A_2, B_2, and W_2 to solve for economic life?

In order to solve this problem, we need a simple way to pass from the variable schedule he experienced to one of 2,000 scheduled hours per year. This can be done by developing two curves, one for cumulative scheduled hours versus cumulative operating hours and another for cumulative variable costs versus cumulative operating hours.

The first step is to develop the statistics of the actual cost records. This can best be done in tabular form. Table 6-2 shows the development of this information. Columns 1, 2, 4, and 6 are the given data. Columns 3, 5, and 8 are the cumulative amounts obtained by adding up the amounts of previous years. Column 7 is a list of the annual variable costs obtained by subtracting the fixed costs ($300) from the previous column. Figure 6-2 shows a plot of the actual M & O costs as they were incurred. Obviously, they do not look as annual costs should look. The fifth-year cost is less than the third and fourth and the second-year cost is less than the first. One would naturally expect the cost to increase annually as the machine aged, but the number of operating hours has obscured this cost rise.

The second step is to construct curves of cumulative hours and costs. The data in Table 6-2 are used to draw Figs. 6-3 and 6-4. This direct approach shortens the solution by sidestepping availability and going directly from scheduled hours to operating hours. If availability is known along with operating hours, scheduled hours can be determined by the relationship given in a previous section of this chapter.

The final step is to pass from actual hours and costs to the selected schedule of 2,000 hours/year. This schedule could just as easily be 1,600 or 2,165 hours: the technique remains the

Table 6-2 Actual hours and cost of machine

Year (1)	Scheduled hours (2)	Cumulative scheduled hours (3) = Σ(2)	Operating hours (4)	Cumulative operating hours (5) = Σ(4)	Annual M & O costs (6)	Annual variable costs (7) = (6) − $300	Cumulative variable costs (8) = Σ(7)
1	2,210	2,210	2,170	2,170	1,430	1,130	1,130
2	1,850	4,060	1,710	3,880	1,320	1,020	2,150
3	2,200	6,260	1,930	5,810	1,660	1,360	3,570
4	2,130	8,390	1,840	7,650	1,800	1,500	5,070
5	1,610	10,000	1,390	9,040	1,510	1,210	6,280

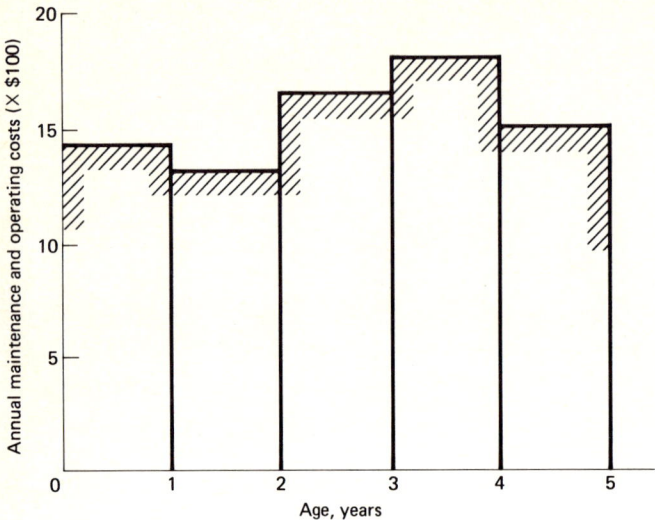

Fig. 6-2 Actual annual maintenance and operating costs.

same. Table 6-3 shows the data derived from Figs. 6-3 and 6-4 for the average of 2,000 hours/year. Keep in mind that these are scheduled hours and not operating hours. Columns 1, 2, and 3 are known data. Column 4 shows the cumulative operating hours corresponding to the scheduled hours of column 3 and Fig. 6-3. Column 5 shows the costs related to the operating hours in column 4 and found in Fig. 6-4. Column 6 shows the annual variable costs obtained from column 5. Column 7 shows the M & O costs obtained by adding the fixed costs ($300) to the variable costs.

Figure 6-5 is a plot of the cost data in column 7. By drawing a smooth curve joining the centers of the bars in the graph, the intercept is found to be $1,200. This is the value of A_2. Assume that $B_2 = A_2$ so that the upper limit of the curve is $A_2 + B_2 = \$2,400$. The annual geometric rate of increase of cost W_2 is 0.15. In this way the actual costs related to a variable schedule can be used to determine the projected cost of a schedule of any number of hours per year.

The data of Table 6-3 can also be used to determine a revenue-cost curve. The cumulative operating hours developed in column 4 are used to find the number of hours of operation per year. This number of annual hours of operation multiplied by the rental or use rate will equal the annual revenue derived from the machine.

Table 6-3 Derived hours and costs of machine

Year (1)	Scheduled hours (2)	Cumulative scheduled hours (3) = Σ(2)	Cumulative operating hours (4) ← Fig. 6-3	Cumulative variable costs (5) ← Fig. 6-4	Annual variable costs (6) = (5) − Σ	Annual M & O costs (7) = (6) + $300
1	2,000	2,000	1,960	980	980	1,280
2	2,000	4,000	3,820	2,120	1,140	1,440
3	2,000	6,000	5,600	3,400	1,280	1,580
4	2,000	8,000	7,340	4,790	1,390	1,690
5	2,000	10,000	9,040	6,280	1,490	1,790

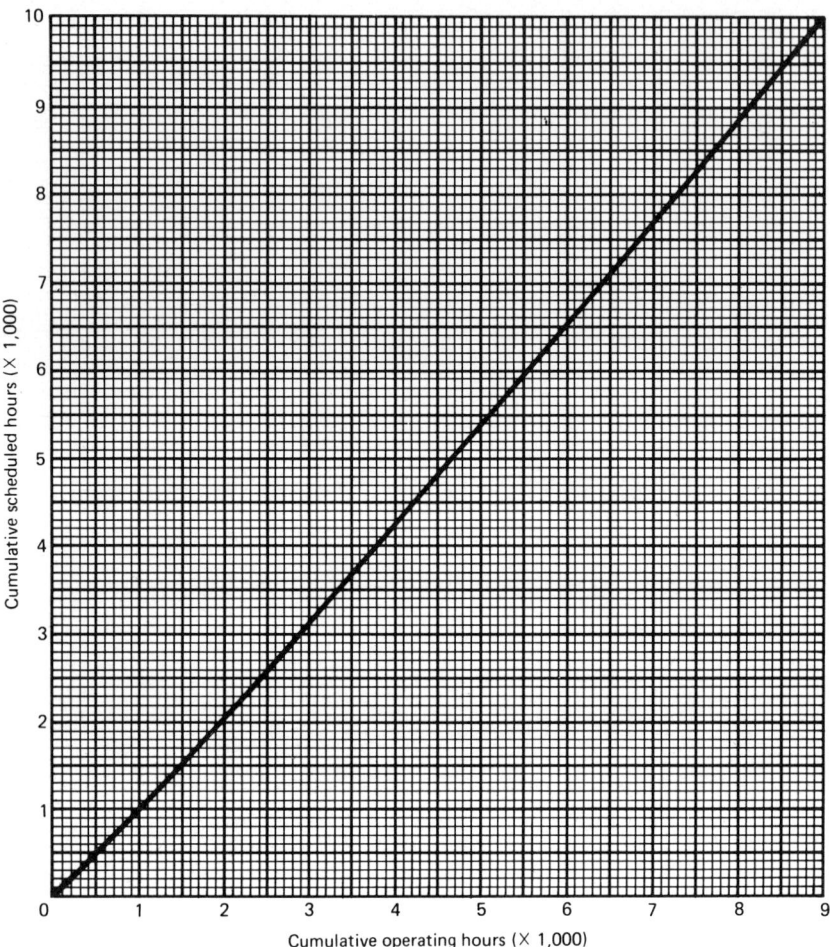

Fig. 6-3 Operating hours versus scheduled hours.

MANAGEMENT REPORTS

The outcome of all this record keeping must be the submission of periodic reports to all levels of management. Usually these reports will be periodic in that they will be tendered at regular intervals depending on the source, purpose, and type of report. Situation reports are those made at a particular time to determine existing facts and are not repeated except on demand.

How often reports should be made is a matter of considerable controversy, expecially between those making them and those receiving them. Certainly, reports should be made no oftener than required to maintain good control of operations. Periodic reports need not always be examined at the time they are

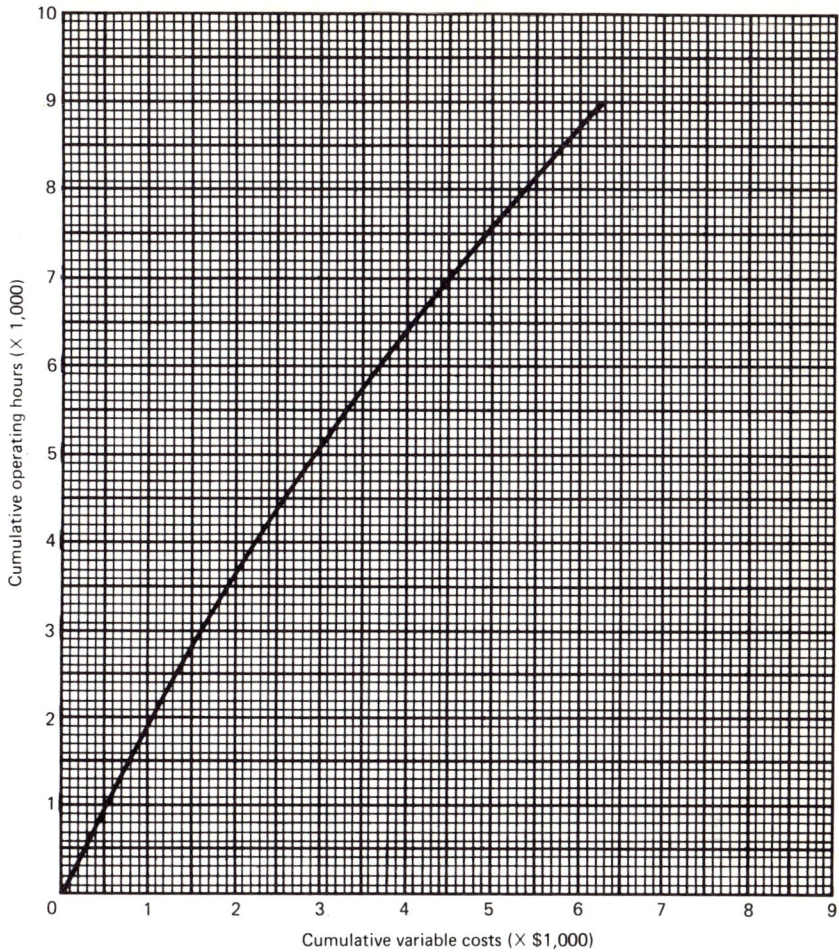

Fig. 6-4 Variable costs versus operating hours.

made. Many are accumulated until a meaningful trend in operations is observable. Reports on production and costs should be rendered weekly on most operations and daily on critical ones. This is easy to accomplish on the computer provided the data can be collected, checked, and entered into the system in a timely fashion.

Not every report need be brought to the attention of management at the time it is rendered. The key to successful reporting is exception reporting. For example, fuel costs per hour might be figured for 100 similar machines. Only those which deviate from some standard by a given amount, say 20 percent, should be brought to the equipment supervisor's attention. He

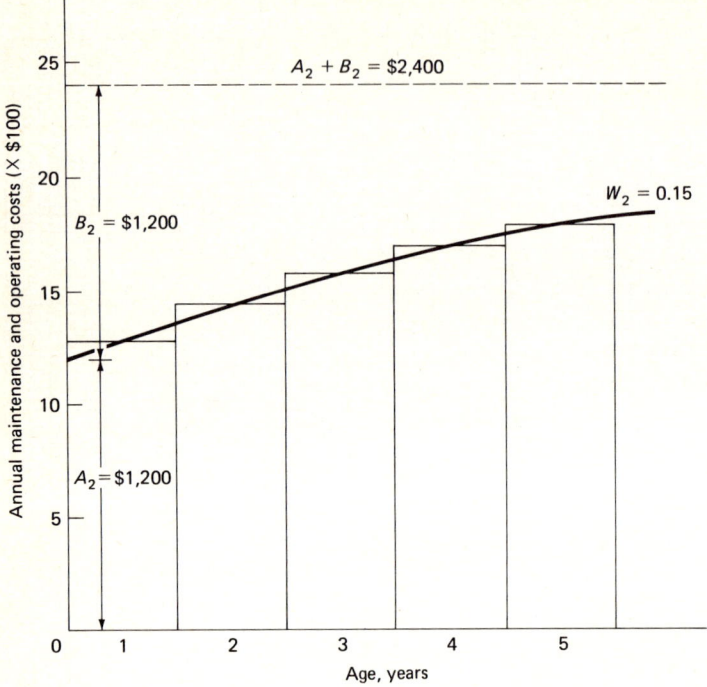

Fig. 6-5 Annual maintenance and operating costs for 2,000 scheduled hours per year.

may then make a cursory examination and decide whether further investigation is warranted.

All reports of costs and hours should be routed only to the management levels required for action and review. Each operator should be kept informed on a continuing basis of the cost of his machine. Each foreman should see the costs of the machines in his spread. Composite reports should be seen by the equipment superintendent at monthly, quarterly, and annual intervals. Those related to policy should be seen by the top level of management quarterly and never less than annually.

SUMMARY

All owners of equipment must keep cost, time, and production records of their equipment. The extent of these records will vary with the type of work, competition, and pressures on the contractor to improve his efficiency of operation. As a minimum, permanent records of the annual costs of individual machines should be kept in the following categories:

1. Revenue
2. Fixed costs
    a. Depreciation
    b. Interest, insurance, and taxes
    c. Storage and security
    d. Fees, licenses, and fines
    e. Moving
    f. Overhead
3. Variable costs
    a. Maintenance
    b. Field repairs
    c. Fuel
    d. Tires or tracks
    e. Operating labor
    f. Field supervision
    g. Major repairs
    h. Overhaul

Times should be entered on all machines individually to maintain annual totals of the following:

1. Shift hours
2. Idle time
3. Scheduled hours
4. Downtime
5. Operating hours

A permanent record should also be kept of production by years. It is when these permanent records are available for each machine that they can be used to analyze performance, determine economic life and replacement timing, and obtain the true cost of a machine over its lifetime of useful service.

REFERENCES

1. AIA/AGC/CSI/CMSI: "Uniform System for Construction Specifications, Data Filing and Cost Accounting," 1966.
2. ASCE: "Construction Cost Control," 1951.

3. Brock, Dan S.: "Cost Accounting Manual for Highway Contractors," American Road Builders Association, Washington, 1971.
4. Cox, Elmer A.: Information Needs for Controlling Equipment Costs, *Highw. Res. Board Highw. Res. Rec.* 278, Washington, 1969.
5. Hackney, John W.: "Control and Management of Capital Projects," Wiley, New York, 1965.
6. Lewis, B. T., and J. P. Marron: "Management of Vehicular Operations and Maintenance," Rider, New York, 1965.
7. Stewart, R. C.: Motor Vehicle Fleet Management, *Am. Public Works Assoc. Res. Found. Spec. Rep.* 37, Chicago, 1970.
8. Walker, Frank R.: "Practical Accounting and Cost Keeping for Contractors," 5th ed., Walker, Chicago, 1961.
9. Wolkstein, H. W.: "Accounting Methods and Controls for the Construction Industry," Prentice-Hall, Englewood Cliffs, N.J., 1967.

7
Equipment Financing

USE OF EQUIPMENT

The purpose of equipment financing is to obtain equipment for use. It is often stated that profits are derived from the use of equipment and not from owning it. There are three basic methods of financing equipment for use, purchasing, leasing, or renting. There are at least several different ways of accomplishing each of the above methods and sometimes combinations of them. Which of the methods is used to acquire equipment for use will depend on several factors which will be discussed in the ensuing pages. Methods of comparing the economic advantages of buying, leasing, or renting will also be examined to help decide which method is to the best advantage of the construction contractor.

Many years ago, the only acceptable method of acquiring machines for use was by direct purchase. Although deals were probably consummated for the rental or short-term use of equipment, it was not common to lease or rent. The leasing business as we know it today has enjoyed its fastest growth since the end of World War II. It began to mature in the late sixties and has now become a common and acceptable method of acquiring equipment for use in business.

There are several reasons for the high rate of growth of leasing after the war. In the manufacturing business, most machine tools had become worn and obsolete as a result of heavy production during the war. Capital was needed to replace them. A dearth of working capital in the companies themselves pointed the way to outside capital which might be made available through leasing. Financial institutions such as banks, savings associations, insurance companies, and the like found a ready and lucrative market in leasing everything for which a demand exexisted. Today "if you can buy, you can lease it" is a common expression.

Another compelling reason for leasing became obvious with the advent of computers. Computer technology has advanced so rapidly in the past 2 decades that most computers became obsolete in 3 or 4 years. For many users, the obvious answer to computer obsolescence was short-term leasing rather than buying.

It has been estimated that the total original value of equipment leased in 1966 was $1.35 billion[1] and that by 1975 it will be about $7 billion.[2] Evidently, the leasing industry is enjoying a healthy growth with an attractive profit for those engaged in it, but it is also attractive from the user's point of view, and this will be discussed later.

Renting has not grown so much as leasing. No figures are available for the magnitude of the rental business since so many people are engaged in it as a secondary enterprise. Nearly all contractors are willing and anxious to rent their idle equipment—at the right price. Rental here is meant to imply the short-term use of equipment as measured in hours or months. Although rental rates are much higher than ownership costs, they are justified by the uncertainty and short-term use of the equipment. Renting serves a very useful purpose for the contractor under certain conditions. These will be discussed later in the chapter.

---

[1] Ray Schuster, The Plant Engineer and Equipment Leasing, *Plant Eng.*, May 1967.
[2] R. R. MacNabb, Leasing of Industrial Equipment, *Leasing of Industrial Equipment: A MAPI Symposium*, Machinery & Allied Products Institute, Washington, 1965.

EQUIPMENT FINANCING

PURCHASE OF EQUIPMENT

Purchase of equipment implies ownership; i.e., the purchaser acquires title to the machine and an equity in it. He is entitled to special treatment under the tax laws which a nonowner, or lessee, does not acquire. For example, he is entitled to deduct depreciation on the machine as an expense of doing business. He may deduct interest on his equipment loan if he has one. He may also receive investment credit if he is entitled to it. The fact that he owns the machine obliges him to pay fees and property taxes and to fulfill the other obligations of ownership.

Direct ownership of equipment has the following advantages:

1. It allows the owner to utilize depreciation and interest on equipment loans as tax-deductible business expenses.
2. It improves the psychology of maintenance and pride of operation through direct ownership.
3. It gives the owner complete freedom to use his equipment as he wishes and to dispose of it whenever and wherever he finds advantage.
4. It assures the owner of any investment credit which may be acquired with the purchase of the machine.
5. The owner benefits directly from wise disposal and salvage value.

There are several methods of purchasing equipment: outright purchase, bank loan, rental with option to buy, and lease purchase. The first two methods will be discussed here and the last two under renting and leasing.

OUTRIGHT PURCHASE

This is often called direct purchase and is the oldest and perhaps safest method of acquiring equipment. The buyer pays his cash for the machine, and the seller releases his title to the buyer. It is the safest method because the transaction is clean: there are no further payments on the part of the buyer, and the seller has only his warranty to fulfill. The most difficult part for the buyer is to have the cash available, since it usually must come from otherwise available working capital. There are no interest payments, concealed financial obligations, or hidden arrangements that occur in the other methods of financing.

The computation of capital costs is simple; they are all paid at the beginning of the life of the machine. The owner has a choice of the type of depreciation to amortize the cost of his machine. He has absolute control of the use and disposition of the machine because it is his property. He can sell it, trade it, or wear it into the ground without having to answer to any creditor. Many people like this feeling. The owner with cash for a purchase is usually able to get better discounts and more benefits because of his strong financial position in the deal. He does not have to choose a particular machine because it is the only one he can get financed. He does not have to accept any inferior machine because he is unable to get approval from a lending institution for the machine of his choice. He has complete independence and autonomy in selection and timing of his purchase. For those reasons, outright purchase will continue to be a popular method of financing among contractors with ready cash.

BANK LOANS

Bank loans and loans from financial institutions are often used to purchase equipment. The owner can take depreciation on his machine since he holds the title to it. He cannot deduct his payments on the principal of the loan as a business expense since he already is writing off the depreciation expense, but he can deduct interest as an expense in computing his taxes. Under the Armed Services Procurement Regulations (ASPR), he can use depreciation expense as a collectible item in a cost-plus-fixed-fee (CPFF) contract with the federal government, but he cannot get reimbursed for the interest expense on his equipment loan. This rule makes leasing favored over purchase by loan on CPFF contracts.

On the bank loan, the lending institution may require a down payment of 20 to 30 percent or the maintenance of a compensating balance of like amount. In other words, they are not willing to lend more than 70 to 80 percent of the cost of the machine, exclusive of taxes, freight, and installation costs. The machine usually is the collateral for the loan; i.e., it becomes the property of the bank in case of default on the loan. A bank likes to deal with a customer who has money on deposit in the bank. Depending on the financial condition of the borrower, he may be required to maintain a credit balance of 20 percent or more in order to guarantee payments on the loan. If a compensating balance of 20 percent is required, the bank may lend the full amount, in which case interest is paid on the full amount of the loan, thus raising the effective interest rate by $\frac{1}{4}$ ($i/0.8 = 1.25i$). In other words, a quoted loan interest rate of 8 percent becomes 10 percent if a 20 percent compensating balance is required. The term of a bank loan is usually 2 to 4 years, certainly no greater

EQUIPMENT FINANCING

than about 75 or 80 percent of the useful life of the equipment. The bank needs to be sure that the owner always has enough equity in the machine to make abandonment undesirable.

The two methods generally used in computing interest for bank loans are *simple interest* and *add-on interest*. Let us look at the details of computing interest with these methods.

Simple interest This is really the nominal rate of compound interest, computed each time a payment is made. It is frequently called the *annual percentage rate*. If payments are made monthly at a simple interest rate of 9 percent, then the interest payment each month will be 0.09/12 times the borrowed amount. When equal monthly payments are made, one part is the interest and the remainder goes toward the principal to retire the loan.

Example Assume that a contractor borrows $10,000 at 9 percent for 4 years. (Payments start July 1.) What will his monthly payments be? What will his balance be at the end of 6 months? How much interest can he deduct as a tax expense?

Solution Look up the capital-recovery factor in a set of interest tables for 0.0075 [(9 percent)/12] with repayment in 48 periods. This factor times $10,000 is the monthly payment and equals $248.85. Table 7-1 shows the results of the computations. It will be observed that the interest paid each month gradually decreases as the principal decreases. This is favorable to the taxpayer since the interest can be deducted as an expense and the largest interest payments occur soonest. Because this money has a time value, it is worth more than equal monthly payments in interest when discounted back to the present. It favors the taxpayer the same way as accelerated depreciation, since it approaches zero at the end of the fourth year as the loan is amortized. It will be seen in Table 7-1 that the balance due at the end of 6 months is $8,937.15 and the total amount of interest deductible as an expense is $430.25.

Simple interest is more difficult for the average buyer to

Table 7-1 Payment on bank loan with simple interest

Payment		Balance		Interest	Principal
Date	Amount	Old	New		
(1)	(2)	(3)	(4) = (3)-(6)	(5) = (3) × 0.0075	(6) = (2)-(5)
July 1	$ 248.85	$10,000.00	$9,826.15	$ 75.00	$ 173.85
Aug. 1	248.85	9,826.15	9,651.00	73.70	175.15
Sept.1	248.85	9,651.00	9,474.53	72.38	176.47
Oct. 1	248.85	9,474.53	9,296.74	71.06	177.79
Nov. 1	248.85	9,296.74	9,117.62	69.73	179.12
Dec. 1	248.85	9,117.62	8,937.15	68.38	180.47
Total	$1,493.10			$430.25	$1,062.85

comprehend since it requires some knowledge of compound interest (unless one is to rely entirely on a set of interest tables). A more direct, simple, and hence more common method of computing interest is the add-on interest method.

Add-on interest This is commonly used in computing car loans. Because of the method of computing it, the rate is deceptive when compared to simple interest calculations. Interest computed by add-on is approximately double the simple interest at the same rate.

Add-on interest is figured by applying the stated rate of interest to the total amount borrowed for each year of the term. For example, if $4,000 is borrowed for 3 years at 5 percent add-on, the total interest paid is $3 \times 5\% \times \$4,000 = \$600$. If this loan were to be repaid in equal monthly installments, the monthly payments would be $(1.15 \times \$4,000)/36 = \$127.78$. This would repay the $4,000 loan plus $600 interest in 3 years. In order to see how this method works, let us turn to another example.

Example Using the same monthly payments found in the previous example with 9 percent simple interest, determine the add-on rate which equals the simple interest rate. Now what will be the balance and total interest paid at the end of the first 6 months?

Solution The total amount of money paid over 48 months is $48 \times \$248.85 = \$11,944.80$. Since the principal was $10,000, the remaining $1,944.80 will be the interest payment. Each year the borrower will have paid $\$1,944.80/4 = \$486.20$ in interest. The add-on rate will be $\$486.20/\$10,000 = 0.0486$, or 4.86 percent. Table 7-2 shows how the balance and interest are worked out for the first 6 months. Notice that although the monthly payments are the same as before, the interest deductions are almost half those in the previous example ($243.12 compared with $430.25). It is obvious that the present worth of interest to the taxpayer in the first calculation in Table 7-1 will be considerably greater than that computed by add-on in Table 7-2.

In summary, it may be said that bank loans used in the purchase of equipment have the following characteristics:

1. A down payment of 20 to 30 percent is required.
2. If no down payment is made, the bank will usually require a compensating balance of like amount, and interest is figured on the total amount lent.
3. Interest on the loan is figured as:
   a. Simple interest, which is determined as a fixed percentage of the balance at the time of each payment.

Table 7-2 Payment on bank loan with add-on interest

Date	Payment Amount	Balance Old	Balance New	Interest	Principal
(1)	(2)	(3)	(4) = (3) − (6)	(5) = $10,000 × $\frac{0.04862}{12}$	(6) = (2) − (5)
July 1	$ 248.85	$10,000.00	$9,791.67	$ 40.52	$ 208.33
Aug. 1	248.85	9,791.67	9,583.34	40.52	208.33
Sept. 1	248.85	9,583.34	9,375.01	40.52	208.33
Oct. 1	248.85	9,375.01	9,166.68	40.52	208.33
Nov. 1	248.85	9,166.68	8,958.35	40.52	208.33
Dec. 1	248.85	8,958.35	8,750.02	40.52	208.33
Total	$1,493.10			$243.12	$1,249.98

b. Add-on interest, which is determined as a fixed percentage of the total amount of the loan.
4. For equal total interest on a loan, the add-on rate will be approximately one-half the simple interest rate.
5. The equipment generally becomes the collateral to the loan (or guarantee that it will be paid).

LEASING EQUIPMENT

A lease is a contract between the owner (lessor) of a machine and its user (lessee). For the privilege and obligations of using the machine, the lessee pays rent periodically to the lessor. The period of rental payments may be any designated time period: weekly, monthly, quarterly, or annually. Monthly rental payments are most common. Leases are usually written for a year or more. Many leasing companies will not execute a lease on a piece of equipment worth less than $25,000, preferring to lease more expensive, long-lived equipment to improve the stability of their business.

The advantages claimed for leasing are many, depending on whose point of view is considered. From the contractor's point of view, the following seem to be the most important:

1. The lease provides another source of credit; it enlarges the credit pool.
2. It releases working capital by providing up to 100 percent financing for new or used equipment.
3. By its tax advantages, it reduces the contractor's tax obligations.
4. It creates a favorable cash flow by paying equipment expenses as they accrue rather than in advance.
5. It improves the contractor's financial ratios.
6. It improves the contractor's financial position in executing CPFF contracts with the federal agencies.
7. It gives the small contractor more leverage in getting warranties and other obligations of the manufacturer fulfilled.
8. It provides an opportunity for the small contractor to take advantage of the large volume use of the lessor in obtaining lower prices for fuel, tires, and other supplies.
9. It enables the contractor to utilize the expertise of the engineering staff of the lessor for guidance in the selection of equipment and its maintenance and management.

CONDITIONS OF LEASES

There are a multiplicity of conditions attached to leasing contracts, depending on the amount involved and who the parties

EQUIPMENT FINANCING 125

are. There are usually three and sometimes four parties involved in the lease: the manufacturer, a lending institution, a lease broker, and the user. The lease broker is optional and serves the same function as an insurance broker. It is he who often puts the deal together.

The term of the lease is usually from 2 to 8 years for equipment. Lessors, like banks making loans, want to be sure of the lessee's financial condition and make the same sort of credit investigation. Once the lessee's good credit has been established, the lessor gives him a free hand to choose the make and model of the machine which will suit him best. When the machine is delivered, the lessor passes all factory service agreements, inspections, and warranties along to the lessee.

Rental payments are made in a variety of ways, usually tailored to the convenience of the lessee. These periodic payments are usually made monthly, sometimes yearly, but can be adjusted to any period agreed on by both parties. Some of the different types of payment schemes used are:

1. Equal monthly or annual payments
2. Skip payments
3. Monthly payments declining annually

Lease payments Probably the most common lease payments are those made monthly. Payments are sometimes required in advance, sometimes in arrears. In leases they are usually paid in advance. There will be a small difference in the discount factors of payments in advance from those in arrears because of the 1-month difference in timing. The discount factor for 12 monthly payments at 6 percent in arrears is 0.9682 while the factor for 12 monthly payments at the same rate in advance is 0.9731. The discount factor for the payments in advance will always be greater because the money is returned 1 month sooner.

When a lease contract is signed by both parties, the lessee becomes obliged to make payments in accordance with the terms of the lease. He is "locked in" on the payments for the term of the lease unless there is an escape clause. Since financial penalty is assessed whenever the lessee terminates the lease early, leases must be treated as long-term obligations.

Payments can be calculated by using either simple interest or add-on interest, as described previously. In the determination of monthly payments by add-on, the amount of the payment can be calculated by the following equation

$$m_a = \frac{(1 + at)S}{12t}$$

where m_a = monthly payment
S = amount borrowed
a = add-on interest rate as a decimal fraction
t = term in years

Example If $5,500 is borrowed at an add-on interest rate of 7 percent for 4 years, what will the monthly payment be?

Solution

$$m_a = \frac{(1+at)S}{12t}$$

$1 + at = 1 + (4 \times 0.07) = 1.28$
$S = \$5,500$
$12t = 12 \times 4 = 48$

$$m_a = \frac{1.28 \times \$5,500}{48} = \$146.67$$

Payments will be $146.67 per month.

When simple interest is used, the calculations are best performed by an electronic calculator. A set of interest tables or the equation below can be used to obtain the capital recovery factor.

$$i = \frac{s}{12}$$

and

$$\text{crf} = \frac{i(1+i)^n}{(1+i)^n - 1}$$

$$m_i = \text{crf} \times S$$

where crf = capital recovery factor
i = interest rate as a decimal fraction
n = number of months
s = nominal annual rate of simple interest
m_i = monthly payment
S = amount borrowed

Example If $5,500 is borrowed at a simple interest rate of 12 percent for 4 years, what will the monthly payment be?

Solution

$$i = \frac{s}{12} = \frac{0.12}{12} = 0.01$$

$n = 48$
$(1+i)^n = (1.01)^{48} = 1.6122$

$$\text{crf} = \frac{i(1+i)^n}{(1+i)^n - 1} = \frac{0.01 \times 1.6122}{1.6122 - 1} = \frac{0.016122}{0.6122} = 0.02633$$
$$m_1 = \text{crf} \times S = 0.02633 \times \$5{,}500 = \$144.82$$

Payments will be $144.82 per month.

Table 7-3 shows the monthly payments on leases at various annual interest rates per $1,000. To check the problem above, locate the nominal annual rate of 12 percent and go across to the column for a 4-year term. The amount shown is $26.33 per $1,000. The monthly payment will be

$$m_1 = \$26.33 \times \frac{\$5{,}500}{\$1{,}000} = \$144.82$$

Skip-payment plans are figured for contractors who like to make their payments during the working season when they have income. Payments are worked out for the working season so that none are made in December, January, February, and sometimes March. They are usually figured by add-on interest, so the computation is made by dividing the annual amount of principal and interest by 8 or 9 instead of 12, as in the case of equal monthly payments.

Example Suppose a contractor desired a skip-payment plan for $5,500 at 7 percent add-on for 4 years. He does not want to make payments during the three coldest months in the winter. How much will he pay each of the other nine months?

Table 7-3 Monthly payments on lease at various rates and periods per $1,000

Nominal annual rate,%	Period, years					
	2	3	4	5	6	7
5	43.87	29.97	23.03	18.87	16.10	14.13
6	44.32	30.42	23.49	19.33	16.57	14.61
7	44.77	30.88	23.95	19.80	17.05	15.09
8	45.22	31.33	24.41	20.27	17.53	15.58
9	45.68	31.80	24.89	20.76	18.03	16.09
10	46.16	32.28	25.37	21.25	18.53	16.61
11	46.60	32.73	25.84	21.74	19.03	17.12
12	47.07	33.21	26.33	22.24	19.55	17.65
$13\frac{1}{2}$	47.78	33.94	27.08	23.01	20.34	18.46
15	48.49	34.67	27.83	23.79	21.15	19.30
18	49.92	36.15	29.38	25.29	22.81	21.02

Solution The annual payment will be $\{[1 + (0.07 \times 4)] \times \$5,500\}/4 = (1.28 \times \$5,500)/4 = \$1,760$. The monthly payment will $= \$1,760/9 = \$195.56$ for each of the 9 months of the working season.

There are several schemes for making monthly payments which decline annually. They are commonly based on the sum-of-years-digits (SOYD) depreciation method. One such plan is the U.S. Leasing Corporation's SOYD. Another is the C.I.T. Corporation's PAYD (pay as you depreciate), described here.

In the C.I.T. plan, called PAYD, equal monthly payments are made which decline annually as the machine ages. Interest is figured by add-on, and payments are made to retire the total amount borrowed plus interest. Table 7-4 shows the percent monthly payment by year for 3 to 6 years. A comparison of SOYD depreciation and the PAYD plan shows the annual amount paid on a 5-year term:

Year	SOYD	PAYD
1	0.333	0.336
2	0.267	0.264
3	0.200	0.204
4	0.133	0.132
5	0.067	0.064
Total	1.000	1.000

Monthly payments are easily calculated by the factors shown in Table 7-4 once the amount and terms of the lease are known.

Example A contractor wants to finance a machine by PAYD on a 4-year term. The machine costs \$8,000, and interest will be figured at 6 percent add-on.

Solution

Total amount $= [1 + (4 \times 0.06)] \times \$8,000 = 1.24 \times \$8,000 = \$9,920$

Monthly payments will be (see Table 7-4):

first year: $0.0340 \times \$9,920 = \$337.28$
second year: $0.0250 \times \$9,920 = \$248.00$
third year: $0.0160 \times \$9,920 = \$158.72$
fourth year: $0.0083 \times \$9,920 = \$82.34$

Amount of lease While banks, equipment distributors, and loan companies are willing to lend small amounts (usually at higher interest rates), the large lessors seek larger commitments. Citicorp Leasing, Inc., of New York, has set a minimum of \$25,000

Table 7-4 Monthly payment on PAYD plan as a percentage of the total amount

Year	Term, years			
	3	4	5	6
1	4.20	3.40	2.80	2.40
2	2.80	2.50	2.20	2.00
3	1.33	1.60	1.70	1.60
4		0.83	1.10	1.20
5			0.53	0.80
6				0.33

with terms ranging from 3 to 10 years. Corporations of this size are willing to negotiate a master lease to cover the total equipment on a job. A master lease would have an upper limit of, say, $10 million. The rate of interest would be fixed, as would the term of the lease. As the contractor mobilized for the job, equipment would be purchased and covered by the master lease without negotiating a new lease for each machine.

TYPES OF LEASES

There are many types of leases ranging all the way from the base lease to the full-service lease. Lessors include national banks, equipment manufacturers, self-financed professional leasing companies, and leasing brokers. The types of leases negotiated by these lessors depend on their capabilities in the leasing business. Those who are strictly money merchants will negotiate bare leases which provide nothing more than the financing. A larger firm which specializes in leasing will have a full staff of engineers and managers, capable of providing complete guidance, maintenance, and everything except a use for the equipment. In the construction-equipment field, the bare lease is the most common. In electronic data-processing equipment, parcel and delivery service, and service-oriented business, the full-service lease is often used. The full-service lease furnishes everything: the equipment, full maintenance and repair, operating supplies, accounting, and management, including standby equipment. The lessee usually provides only the driver and the operating schedule.

In the bare lease, the lessee usually buys his own supplies and parts. At times, however, the purchasing power of the lessor is used to obtain sizable discounts on tires, batteries, parts, and other consumables. Where the full-service lease is used, all these items are procured directly by the lessor.

Some leases contain a provision for the lessee to acquire title to the equipment if he should later decide to buy instead of lease. These contracts are known as leases with option to buy. The purchase option is usually one that allows the lessee to acquire title by applying some of the rents already paid to the purchase price of the equipment.

A typical lease with option to buy might include the following provisions. A crane worth $100,000 and with a useful life of 10 years could be leased with a guaranteed salvage value of 3 percent in 7 years. The term of the lease would be 7 years with 84 equal monthly payments in advance. The monthly rent would be $1,600 ($16 per $1,000 = running yield 9.2 percent). At the end of the 7-year term, the lessee would be given three options:

1. Return crane to lessor
2. Buy at fair market value
3. Renew lease for 1 year at a time at annual rent equal to 1 month's previous rent of $1,600

Leases with option to buy must be carefully scrutinized by the lessee. He should also obtain the advice of his accountant and tax lawyer. This type of lease is often interpreted by the Internal Revenue Service as a conditional sales contract, which precludes deduction of rental payments as expenses.

TAX IMPLICATIONS

Rents are usually deductible as an expense of doing business. This is one of the principal advantages of leasing. Because money has a time value, near payments are worth more as tax deductions than distant payments; concomitantly, near costs are relatively greater than far costs when discounted. For that reason, with comparable nominal interest rates, the present worth of costs on a lease plan with equal monthly payments will be less than those based on a SOYD plan. The effect of income and corporation taxes on the economics of ownership is sufficient to justify a careful assessment of their impact.

Rental expense deducted from gross income is a powerful factor in economic analysis in favor of leasing. Conversely, depreciation deductions and gain on sale have considerable weight in direct ownership. Investment credit always goes to the owner of a machine. With direct purchase there is no doubt who is entitled to the credit. In leasing, the lessor usually passes it on

EQUIPMENT FINANCING

to the lessee, either directly or in the form of reduced payments. Taxes, fees, insurance, and other ownership costs are passed on to the lessee by the lessor, and these, too, are deductible expenses against income and corporation taxes.

Whether the lease is a true lease or is in fact a conditional sales contract is the most controversial factor in leasing. If the contract is interpreted as a conditional sales contract, lease payments are not tax-deductible and the equipment must be capitalized by deductions for depreciation.

Revenue Ruling 55-540 sheds considerable light on the interpretation of Internal Revenue in separating leases from conditional sales contracts. Section 4.01 states:

> Whether an agreement, which in form is a lease, is in substance a conditional sales contract depends upon the intent of the parties as evidenced by the agreement, read in the light of the facts and circumstances existing at the time the agreement was executed.

The ruling goes on to say that a transaction will generally be treated as a sale or purchase rather than a lease if one or more of the following conditions are present:

1. Some of the periodic payments are applied to some equity to be acquired by the lessee.
2. The lessee acquires title upon the payment of stated "rentals" which he is required to make.
3. The lessee pays an inordinately large amount of the total sum required for transfer of title in a short time.
4. The agreed "rental" payments are greatly in excess of the current fair rental value.
5. The property may be acquired under an option to buy at a low price in relation to the actual value of the equipment at the time the option is exercised.
6. A portion of the payments is specifically designated as interest or recognizable as such.
7. Title is acquired upon the payment of a total amount (rentals plus option price) which approximates the price plus carrying charges and interest which would have been paid if the equipment had been purchased in the first place.

It may be inferred from the publication of the various revenue rulings on the subject that there has been considerable hassling over it. A true lease, to be safe, should have the following characteristics:

1. There should be no option to buy.
2. Equipment should have residual value at the end of the lease.
3. Lease payments should approximate rental rates on the current market for similar equipment.

If there is a purchase option, the fair market value should be 5 to 25 percent to avoid suspicion. A 1 percent salvage value would be suspect. Token lease-renewal options are also suspicious. The renewal rate may be lower than the regular payments, but token renewal amounts indicate that the machine has already been purchased.

RENTING EQUIPMENT

Renting equipment, in the modern sense, is the short-term use of equipment. Legally speaking, rent is a payment for the use of property (equipment) acquired by lease. A rental agreement, however, usually connotes the use of equipment for no more than several months. Rental rates are usually quoted for use by the hour, day, week, or month. Here, the term "renting" will be used in that sense.

One of the biggest concerns of the contractor when he decides to rent equipment for short-term use is where to obtain it and how much rent to pay. A careful canvassing of the local marketplace is required to find equipment which will be available at the time required. Additionally, the prospective customer should have some idea of the prevailing rates. There are several sources of rate information:

1. Comparison of prevailing rates of several local companies
2. AED *Rental Compilation*[1]
3. *Blue Book Rental Rates for Construction Equipment*[2]
4. EGCA Directory "Suggested Equipment Rental Rates"[3]

---

[1]Published by Associated Equipment Distributors, P.O. Box 97724, Chicago 60690.
[2]Published by Equipment Guide-Book Co., 3980 Fabian Way, Palo Alto, Calif. 94303. Also publishes *Red Book of Rental Rates for Off-Highway Trucks and Trailers* and numerous other equipment data.
[3]Published by the Engineering and Grading Contractors Association of California, 2115 Beverly Blvd., Los Angeles 90057.

The AED *Rental Compilation* is a statistical average of rental rates reported annually by about 800 members of the Associated Equipment Distributors, Chicago. They are rates reported from all over the United States of equipment in good operating condition (equal to 40 percent new condition or better). Equipment is classified generally by type (air, compaction, concrete, etc.) and size (capacity, horsepower, weight, etc.). There is no differentiation by make or model, although a list of serial numbers is contained in the second half of the book to assist in identifying the equipment. Figure 7-1 shows a page from the AED rating book. These are the rental rates published periodically in *Engineering News-Record*.

The *Blue Book Rental Rates for Construction Equipment*, published and updated continuously by the Equipment Guide-Book Co., of Palo Alto, are national statistical averages reported by numerous distributors, equipment dealers, contractors, and others who rent equipment. Rates are reported by type and size. In the more popular types, such as tractors and earth-moving equipment, the breakdown is more complete by make and model. Figure 7-2 shows a page from the *Rental Rate Blue Book*. At the front of each type section, a map relates the rental rates to geographic areas. As well as the monthly, weekly, daily, and hourly rates, the book shows the estimated hourly operating cost (excluding operator's wage). This book costs more than the others but is much more comprehensive and complete.

The EGCA "Suggested Equipment Rental Rates" is published annually for the benefit of its membership. They are local rates, abbreviated, and a good example of rates published by the various contractor organizations. Various other agencies publish rental rates, sometimes for guidance and sometimes for the purpose of fixing or limiting rates for contract purposes. Among these agencies are the public utilities commissions, state highway departments, and various federal agencies.

Rental rates will vary considerably, depending on whether they are quoted by the hour, day, week, or month. Naturally, the shorter the period, the higher the rate. As an example, a Caterpillar D8H crawler tractor might be quoted at the following rates:

$ 35/hour
   280/day
 1,120/week
 3,388/month

Daily rates will be about eight times the hourly rate. Weekly

The rental rates and terms set forth in this compilation are for informational purposes only and not to suggest or to influence the rates or conditions of rental of any item of equipment, as this is a matter which must be determined by the lessee and the lessor of the equipment involved.

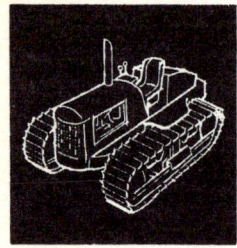

Crawler tractors—without attachments

horsepower from & including	horsepower to & including	one month	one week

Gear drive—diesel engine
Drawbar horsepower

from	to	one month	one week
20	25	$ *	$ *
26	35	800.00	277.00
36	44	*	*
45	59	1049.00	*
60	71	1081.00	*
72	79	1313.00	*
80	91	1439.00	*
92	115	1775.00	*
116	131	2145.00	*
132	—	*	*

Torque converter drive—diesel engine
Net engine horsepower

from	to	one month	one week
115	144	$1683.00	$ *
320	360	*	*

Power shift, torque converter drive—diesel engine
Net engine horsepower

from	to	one month	one week
—	60	$ *	$ *
61	89	1186.00	*
90	119	1717.00	*
120	144	2100.00	*
145	180	2810.00	*
181	250	2970.00	*
251	300	3718.00	*
301	360	*	*
361	—	5588.00	*

Crawler tractors—with bulldozer

horsepower from & including	horsepower to & including	one month	one week

Gear drive—diesel engine
Drawbar horsepower

from	to	one month	one week
20	25	$ *	$ *
26	35	872.00	318.00
36	44	1001.00	338.00
45	59	1122.00	*
60	71	*	*
72	79	1420.00	*
80	91	1630.00	*
92	115	1838.00	591.00
116	131	2156.00	718.00
132	150	2367.00	*
151	170	2525.00	*
171	200	3026.00	*
201	256	3432.00	*
257	—	3875.00	*

Figures shown represent an average of rates reported at the time survey was conducted, and consequently can be considered as current averages only at that time. There are wide ranges of rates in different parts of the country and even in the same trading area—see FOREWORD.

Fig. 7-1 Example page of AED *Rental Compilation. (Associated Equipment Distributors.)*

rates will be about 80 percent of 5 times the daily rate, and monthly rates will be about 55 percent of 22 times the daily rate.

Rates per period will vary according to:

1. Length of working season
2. Climatic conditions and weather
3. Economic conditions
4. Proximity of a market
5. Reputation of user for safety and careful operation
6. Job conditions

MOTOR SCRAPERS

CATERPILLAR

MODEL	Cu. Yds. Capacity	HP	MONTHLY	WEEKLY	DAILY	HOURLY	EST.OPR. COST/HR.
641 PS non-cush. hitch(64 F)	28-38	500	$ 7,145.00	$2,360.00	$590.00	$73.70	$12.30
641 PS cush. hitch (41 M)	28-38	500	7,465.00	2,465.00	615.00	77.00	12.70
650 PS (77 F)	32-44	500	7,920.00	2,615.00	655.00	81.70	13.20
651 PS non-cush. hitch(33 G)	32-44	500	7,765.00	2,560.00	640.00	80.10	13.00
651 PS cush. hitch (44 M)	32-44	500	8,085.00	2,670.00	665.00	83.40	13.40
657 Tw. Eng. PS* (31 G)	32-44	860	6,640.00	2,190.00	550.00	68.50	16.25
657 Tw. Eng. PS. (31 G)	32-44	900	8,380.00	2,765.00	690.00	86.40	17.25
657 Tw. Eng. PS non-cush. hitch(31 G)	32-44	900	9,680.00	3,195.00	800.00	99.85	18.75
657 Tw. Eng. PS cush. hitch (46 M Ser.)	32-44	900	10,000.00	3,300.00	825.00	103.15	19.15
660 PS (77 F Ser.)	40-54	500	8,440.00	2,785.00	695.00	87.05	13.95
666 Tw. Eng. PS* (77 F 113)	40-54	860	7,070.00	2,335.00	585.00	72.90	16.85
666 Tw. Eng. PS (77 F 262)	40-54	900	10,485.00	3,460.00	865.00	108.15	19.70

*Discontinued Model

Rates are National Average - Adjust to Local Conditions.

Fig. 7-2 Example page from *Blue Book Rental Rates for Construction Equipment. (Equipment Guide-Book Co.)*

Because several of the above factors vary with the geographic location, there should be some adjustment for the area where the equipment and job are located.

RATE VARIATION WITH GEOGRAPHIC AREA

Since the length of the working season is materially affected by the location of the rental area, geography becomes one of the most important factors in determining the utilization of equipment. In mountainous areas, where winters are adverse and the working season is short, rates will be higher. In southern California, where winters are mild and there is little rain, rates will be less.

One way to recognize these geographic factors is the method used in the *Blue Book Rental Rates*. Each section of the book has a map of the United States which shows the variation in rate for a particular type. Figure 7-3 shows the area map for section 10, excavators. As might be anticipated, the lowest rates are in places like Florida and Texas and the highest in Alaska, Montana, and the Dakotas.

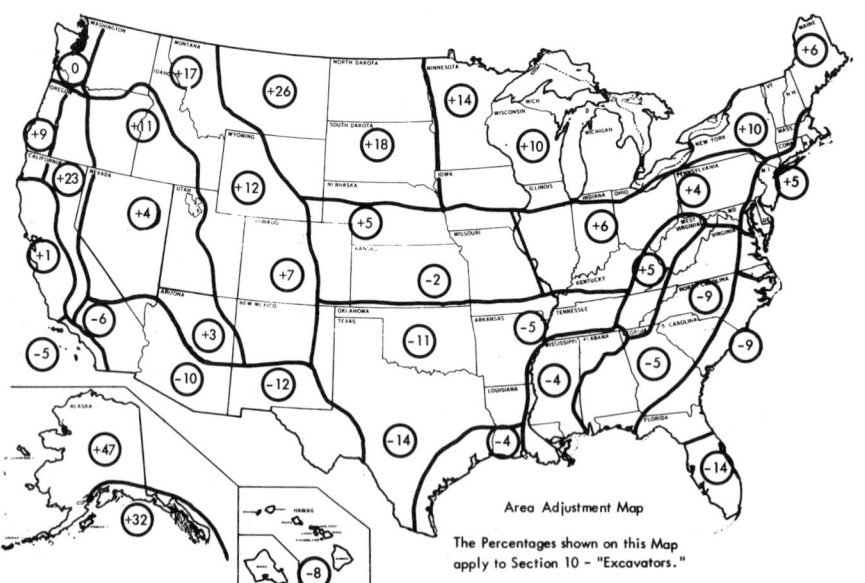

Fig. 7-3 Area map for rate adjustment. *(Equipment Guide-Book Co.)*

EQUIPMENT FINANCING

COMPUTATION OF RENTAL RATES

The computation of rental rates has been a very controversial subject for many years. The principal reason lies in the determination of allowable costs used in determining the rate. At the root of this controversy is the belief of most state highway departments that a contractor's costs are different when he owns a machine from when he rents one. The contractor will not be reimbursed for interest if he owns the machine, but he will pay interest if he rents a machine, and for this he can be reimbursed.

What is an owner entitled to receive for rent? Obviously he must recover *all* costs plus a profit for his effort or he cannot remain in business. Determination of costs is not nearly so difficult as predicting in advance what the utilization will be so you can collect for these costs. The key is the proper estimation of utilization. If many machines are being rented out, the probability of getting a good statistical estimate will be much better than for only one or two machines. The dealer, or large owner, stands a better chance of achieving his statistical-average utilization than a small operator. Contractors state that they usually have to see about 60 percent utilization for a machine to justify purchase. The lessor can achieve this by renting to several different people each year in the life of the machine and therefore make renting cheaper than owning.

More information on establishing rental rates will be found in *Highway Research Board Report* 26[1] and in the front of the AED *Rental Compilation*. Rental rates always include all the ownership costs such as depreciation, interest, taxes, insurance, etc. Often they include maintenance, repair, and operating costs. The shorter the rental period, the more costs the owner will absorb and the higher the rent. When equipment is rented by the hour, it frequently includes the operating labor as well.

GENERAL CONDITIONS OF RENTAL AGREEMENTS

The terms of the agreement should be negotiated by the parties concerned before the user accepts the equipment. The period and rental payment should be clearly stated. The rental period usually begins when the equipment leaves the owner's yard and ends when it is returned. If the equipment is shipped to a distant point, the period may begin and end with the date of the bill of

---

[1] T. S. Dudick and F. I. Ravenscraft, Development of Uniform Procedures for Establishing Construction Equipment Rental Rates, *Highw. Res. Board Nat. Acad. Sci. Nat. Coop. Highw. Res. Prog. Rep.* 26.

lading on which it is shipped out and returned. When equipment is rented on an hourly basis, the lessor usually charges 1 hour's rent for pickup and delivery in the local area. If the distance is great, additional charges will be made for this service.

In figuring normal working time, which is the basis for rental, a day has 8 working hours, a week has 5 working days (40 hours), and a month has 22 working days (176 hours per 30 days). Overtime rates are figured two ways:

1. Any hours in excess of 8 at $\frac{1}{8}$ times the daily rate, $\frac{1}{40}$ times the weekly rate for more than 40 hours in one week, and $\frac{1}{176}$ times the monthly rate for more than 176 hours in a 30-day period.
2. Extra charges computed on a shift basis at 50 percent rate for each additional shift; $1\frac{1}{2}$ times base rate for two shifts, 2 times for three shifts per day.

The owner is responsible for providing the equipment in good mechanical condition. The user is charged with returning it in the same condition less normal wear and tear. What constitutes the latter is often the basis of disagreement between the two parties. Who is responsible for repairs should be clearly spelled out in the agreement. Usually the owner pays for repairs resulting from normal wear and tear while the user pays for others.

Tire costs, especially when working in rock or other adverse conditions, may become a major point of dispute. Who bears these costs will have considerable effect on the rental rate. This will not be so much of a problem with hourly or daily rental as it will be for longer periods. When exceptionally high track wear is anticipated on crawler equipment, the rate will be adjusted upward or provisions should be made in the agreement for a track rebuild before return.

WHY RENT EQUIPMENT

The principal reason for renting equipment rather than leasing or buying is that you cannot utilize it fully over its useful life. It therefore becomes more economical to rent it for the short period you can utilize it fully than to acquire it by other means. By renting equipment, a lessor can increase its utilization and consequently lower the cost for using it.

Other reasons for renting are similar to those for leasing: preservation of capital, improvement of business ratios, no long-term obligation, and better use of working capital. Renting has one asset that leasing does not have, i.e., avoiding obsolescence.

Because renting is a short-term commitment, the contractor is always able to obtain the newest improved machines available. He is able to shift makes and models without building up a large stock of spare parts which might be required to maintain equipment used for a longer term.

Bookkeeping, too, is simplified by equipment rental. The rent is the equipment expense. Because rentals are less than a year in term, there is no problem in deducting rent as an expense of doing business. The machine is almost always returned to the lessor in a few months, thus avoiding the conditional-sale problem often encountered in leasing.

RENT-WITH-OPTION-TO-BUY AGREEMENTS

Rental agreements are sometimes drawn up as rent-with-option-to-buy contracts. The rental rate is usually so high that the user must make a decision in a few months or the deal becomes unfavorable. A typical rent-with-option-to-buy contract might be drawn up as follows:

Rental rate   10 percent of the selling price per month.
Minimum rental period   1 month.
Purchase option   Rental payments for the second through the seventh month shall be applicable to the purchase of the equipment provided the decision to purchase is made within 30 days of the termination of the agreement. Payments for the first month and all months subsequent to the seventh month shall not be applied to purchase.
Other charges   Customer shall pay transportation charges and all other charges such as property, sales, and use taxes, etc.
Monthly payments   Rents are payable monthly and are due on the last day of the month to which they apply.

It is easily discernible that the original value of the equipment plus interest will be paid out in a year. If the machine is returned to the lessor, whatever rents were paid will be deductible tax expenses. If the option to buy is exercised, the machine must be capitalized for tax purposes. A deal of this sort would be best when the buyer is uncertain about the machine's capability and wants to try it before buying. It would almost be a foregone conclusion that if the option to buy were not exercised by the seventh month, the machine would be returned.

COMPARISON OF ALTERNATIVES

Deciding which method of equipment financing to use is best handled by considering the feasibility of the various methods, their economic differences, and their effect on the business as a whole. Several things must be investigated: the influence on business ratios and bondability, the cash flow, and the intrinsic discounted cost of each plan. These can best be studied by using some of the economic tools of analysis described previously.

The decision for short-term rental can be separated from that to lease or buy. Short-term rental should be utilized when the period of utilization is short—less than a year. Any machine that is required for more than a year should be leased or purchased, whichever is more favorable to the contractor. The various methods of leasing and buying can be analyzed by the discounted-cash-flow method to determine which is best in a particular situation. After taking a look at the influence of this decision on business ratios, we shall investigate leasing and buying by finding the present worth of the cost of each plan.

INFLUENCE ON BUSINESS RATIOS

Business ratios are used in many ways to determine the health of an enterprise. Whether you lease or buy can have considerable effect on these ratios. Bondability is too often judged by the contractor's net quick assets, meaning liquid assets which can provide ready cash. As a rule of thumb, it is generally assumed that a contractor is bondable for 10 times his net quick assets. In other words, the outright purchase of a $100,000 machine can reduce his bonding capacity by $1 million.

The use of cash reserves to acquire a piece of equipment will influence business ratios variously, according to the method of acquisition, purchase, bank loan, or lease. In order to illustrate this, assume that a small paving contractor wants to buy a $50,000 machine. Table 7-5 shows the balance sheet before acquisition and after. In this illustration the lease has resulted in the best ratios after acquisition, and this normally is the case.

Table 7-5 Effect on business ratios

	Before acquisition	Cash purchase	Bank loan 20% down	Lease
Current assets	250,000	200,000	240,000	250,000
Fixed assets	150,000	200,000	200,000	150,000
Total	400,000	400,000	440,000	400,000
Liabilities	100,000	100,000	140,000	100,000
Capital	300,000	300,000	300,000	300,000
Total	400,000	400,000	440,000	400,000
Current ratio	2.5:1	2:1	1.7:1	2.5:1
Debt to equity	1:3	1:3	1.4:3	1:3

EQUIPMENT FINANCING

Although a lease is in reality a long-term obligation, it is usually footnoted rather than included in the liabilities.

THE PROBLEM

A small contractor wants to acquire a machine which he expects to use for 5 years. The following facts and assumptions establish the conditions of the problem. He wishes to investigate cash purchase, bank loan, and lease.

Equipment cost $50,000
Lease cost $21/month per $1,000 of cost, equal monthly payments for 5 years in arrears
Bank loan 3-year loan at 6 percent add-on interest with 20 percent down payment, equal monthly payments in arrears
Depreciation Double declining balance for 5-year life; salvage value 15 percent for computing gain on sale
Discount factor Nominal 12 percent annual (1 percent per month for monthly payments)
Investment credit Always to buyer or lessee
Taxes 53 percent of net income

CASH FLOWS AND DISCOUNT FACTORS

Since cash flows are timed differently, two sets of discount factors will be required, those for end-of-year cash flows and those for equal monthly flows in arrears or advance. Since the latter are not shown in most interest tables, it will be necessary to explain first how they can be derived.

The discount factor for monthly payments in arrears can be computed as follows:

Year	Equations
1	$\text{pwf}_{1\text{-}12} = \dfrac{\text{pwf} - i\% - 12}{12}$
2	$\text{pwf}_{13\text{-}24} = \dfrac{(\text{pwf} - i\% - 24) - (\text{pwf} - i\% - 12)}{12}$
3	$\text{pwf}_{25\text{-}36} = \dfrac{(\text{pwf} - i\% - 36) - (\text{pwf} - i\% - 24)}{12}$
4	$\text{pwf}_{37\text{-}48} = \dfrac{(\text{pwf} - i\% - 48) - (\text{pwf} - i\% - 36)}{12}$
5	$\text{pwf}_{49\text{-}60} = \dfrac{(\text{pwf} - i\% - 60) - (\text{pwf} - i\% - 48)}{12}$
n	$\text{pwf} = \dfrac{(\text{pwf} - i\% - 12n) - [\text{pwf} - i\% - 12(n-1)]}{12}$

The discount factor for monthly payments in advance can be computed as follows:

Year	Equation
1	$\text{pwf}_{0\text{-}11} = \dfrac{(\text{pwf} - i\% - 11) + 1}{12}$
2	$\text{pwf}_{12\text{-}23} = \dfrac{(\text{pwf} - i\% - 23) - (\text{pwf} - i\% - 11)}{12}$
3	$\text{pwf}_{24\text{-}35} = \dfrac{(\text{pwf} - i\% - 35) - (\text{pwf} - i\% - 23)}{12}$
4	$\text{pwf}_{36\text{-}47} = \dfrac{(\text{pwf} - i\% - 47) - (\text{pwf} - i\% - 35)}{12}$
5	$\text{pwf}_{48\text{-}59} = \dfrac{(\text{pwf} - i\% - 59) - (\text{pwf} - i\% - 47)}{12}$
n	$\text{pwf} = \dfrac{[\text{pwf} - i\% - (12n - 1)] - \{\text{pwf} - i\% - [12(n - 1) - 1]\}}{12}$

In the above equations, i is the monthly interest rate or the nominal annual rate divided by 12. If the uniform series present-worth factor (pwf) is not available in tabular form, it can readily be computed on an electronic calculator as

$$\text{pwf} = \frac{(1 + i)^n - 1}{i(1 + i)^n}$$

where i = interest rate
n = number of periods

As an example, for an interest rate of 1 percent

$$\text{pwf}_{12} = \frac{(1 + 0.01)^{12} - 1}{0.01(1 + 0.01)^{12}}$$

$(1.01)^{12} = 1.12682$

$$\text{pwf}_{12} = \frac{1.12682 - 1}{0.0112682} = 11.255$$

The discount factor for monthly payments in arrears at 12 percent for the first year would be

$$\text{pwf}_{1\text{-}12} = \frac{\text{pwf} - 1\% - 12}{12} = \frac{11.255}{12} = 0.9379$$

EQUIPMENT FINANCING

Table 7-6 Discount factors for monthly payments in arrears

Year	6%	9%	12%
1	0.9682	0.9529	0.9379
2	0.9120	0.8712	0.8323
3	0.8590	0.7965	0.7387
4	0.8091	0.7282	0.6555
5	0.7621	0.6657	0.5818

and for the second year:

$$\text{pwf}_{13\text{-}24} = \frac{(\text{pwf} - 1\% - 24) - (\text{pwf} - 1\% - 12)}{12}$$

$$= \frac{21.2433 - 11.255}{12} = \frac{9.9882}{12} = 0.8323$$

The product of the pwf and the sum of the monthly payments for the year will be the discounted value of the monthly payments for that year.

As a matter of information, Table 7-6 shows the discount factors for monthly payments in arrears, and Table 7-7 shows the factors for monthly payments in advance.

For use in solving this problem in which interest is quoted at a nominal rate of 12 percent annually, the discount factors in Table 7-8 will be applied to the cash flows.

Let us first consider the cash purchase. Because we are taking double-declining-balance depreciation, there will be a gain on sale. This results from the fast write-off, which brings the book value ($3,888) below the salvage value ($7,500), resulting in a gain on sale of $3,612. Since the investment credit of $2,333 ($\frac{2}{3} \times 0.07 \times \$50{,}000$) will go to the lessee as well as the buyer, it will not be considered. If the lessor refused to pass it on to the lessee, however, it would appear as a cash flow in favor of the les-

Table 7-7 Discount factors for monthly payments in advance

Year	6%	9%	12%
1	0.9731	0.9601	0.9473
2	0.9166	0.8777	0.8407
3	0.8633	0.8024	0.7461
4	0.8132	0.7336	0.6621
5	0.7659	0.6707	0.5876

Table 7-8 Discount factors in example problem

Year	Single payment end of year, discount factor A	Equal monthly payments in arrears, discount factor B
0	1.0000	
1	0.8929	0.9379
2	0.7972	0.8323
3	0.7118	0.7387
4	0.6355	0.6555
5	0.5674	0.5818

sor. Table 7-9 shows the computations for the present worth of the cumulative net cash cost of outright purchase.

Next, let us consider the cost of the machine when purchased by a bank loan. Table 7-10 shows the results of this computation. Again, having used accelerated depreciation, we must pay a gain on sale when the machine is sold at the end of 5 years. Monthly payments on the loan are $1,311.11, or a total of $15,733/year. Since these are monthly payments in arrears, the discount factor B in Table 7-8 will be used for discounting them. End-of-year costs will be discounted by the discount factor A in Table 7-8. In this plan, interest is also deductible as a business expense, so it will be used for a tax credit as well as the depreciation expense.

Finally, consider the lease under the conditions outlined in the problem statement. At the rate of $21 per $1,000, monthly payments will be $1,050 ($21 × 50,000/1,000) in arrears. Table 7-11 shows the computations required for this situation. Note that the monthly payments of $12,600 (12 × $1,050) per year are discounted with discount factor B from Table 7-8 while year-end costs are discounted with discount factor A.

ANALYSIS OF RESULTS

The results of the three computations just performed can best be analyzed by plotting them on a graph and comparing the results. Figure 7-4 is such a plot. Obviously the final result favors the leasing arrangement because less cash flows during the 5-year period and the cumulative cash flow at the end of the fifth year is less than the other two. Cash flow at the end of 5 years is

    Cash purchase $27,345
    Bank loan   23,626
    Lease   23,130

Table 7-9 Computations for cash purchase

Year (1)	Cost (2)	Depreciation DDB at 5 years (3) = 0.4 × BV	Tax credit at 53% (4) = 0.53 × (3)	Discount factor (5)	Present worth of cash flow (6) = (2) & (4) × (5)	Present worth cumulative net cash cost (7) = Σ(6)
0	$50,000			1.0000	$50,000	$50,000
1		$20,000	10,600	0.8929[A]	(9,465)	40,535
2		12,000	6,360	0.7972[A]	(5,070)	35,465
3		7,200	3,816	0.7118[A]	(2,716)	32,749
4		4,320	2,290	0.6355[A]	(1,455)	31,294
5	(7,500)	2,592	(541)	0.5674[A]	(3,949)	27,345
		Gain on sale (3,612)				

Table 7-10 Computations for bank loan at 6% add-on

Year (1)	Cost (2)	Discount factor (3)	Present worth of payments (4) = (2) × (3)	Interest (5) = 0.06 × 40,000	Present worth of interest (6) = (5) × (3)	Depreciation DDB at 5 years (7) = 0.4 × BV
0	$10,000	1.0000	$10,000			$20,000
1	15,733	0.9379[B]	14,756	$2,400	$2,251	12,000
2	15,733	0.8323[B]	13,095	2,400	1,998	7,200
3	15,734	0.7387[B]	11,622	2,400	1,773	4,320
4						2,592
5	(7,500)	0.5674[A]	(4,256)			Gain on sale (3,612)

Year (1)	Discount factor (8)	Present worth depreciation expenses (9) = (7) × (8)	Present worth depreciation & interest (10) = (6) + (9)	Present worth tax credit at 53% (11) = 0.53 × (10)	Present worth of cash flow (12) = (4) + (11)	Present worth cumulative cash cost (13) = Σ(12)
0					10,000	10,000
1	0.8929[A]	$17,858	20,109	10,658	4,098	14,098
2	0.7972[A]	9,566	11,564	6,129	6,966	21,064
3	0.7118[A]	5,125	6,898	3,656	7,966	29,030
4	0.6355[A]	2,745	2,745	1,455	(1,455)	27,575
5	0.5674[A]	(579)	(579)	(307)	(3,949)	23,626

Table 7-11 Computations for lease at $21 per $1,000

Year (1)	Rental payments (2) = 12 × $1,050	Discount factor B (3)	Present worth of rentals (4) = (2) × (3)	Tax credit at 53% (5) = 0.53 × (4)	Discount factor A (6)	Present worth tax credits (7) = (5) × (6)	Present worth cash flow (8) = (4) − (7)	Present worth cumulative cash cost (9) = Σ(8)
0	0						0	0
1	12,600	0.9379	11,818	6,678	0.8929	5,963	5,855	5,855
2	12,600	0.8323	10,487	6,678	0.7972	5,324	5,163	11,018
3	12,600	0.7387	9,308	6,678	0.7118	4,753	4,555	15,573
4	12,600	0.6555	8,259	6,678	0.6355	4,244	4,015	19,588
5	12,600	0.5818	7,331	6,678	0.5674	3,789	3,542	23,130

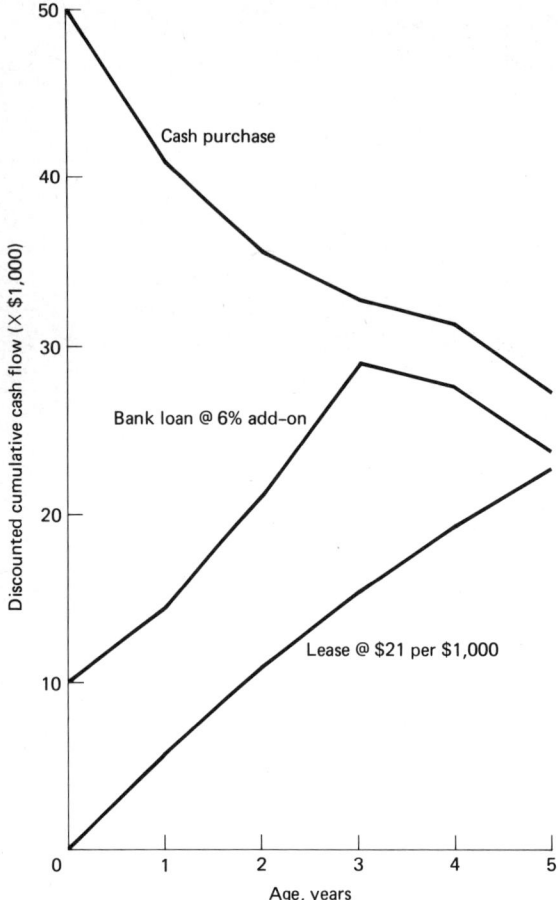

Fig. 7-4 Comparison of equipment-financing plans.

The shape of the curves will remain fairly stable even though some parameters may change. The cash-purchase curve starts at the cash price and gradually descends to the final result. This end result can be manipulated quite a bit by making different assumptions about rate of return and salvage value. The bank-loan curve starts at the down-payment cost, rises to the end of the term of the loan, and then descends to the final cost. The lease, on the other hand, starts at zero and gradually rises to the final cost. Here again, the final cost is subject to manipulation by

EQUIPMENT FINANCING

assuming different rates of return, different rates and types of interest, and different paybacks. Salvage value does not have so much effect on lease plans as it will have on direct-ownership plans such as cash purchase and bank loans.

Other computations will soon show how the final results can be affected by different assumptions which must be made at the time of analysis. These results of final cost are tabulated in Table 7-12. Basic assumptions are those made in the beginning. The first three plans are those which have already been worked out. Others are worked out with the basic parameters changed as noted. The first six plans are all worked out at a 12 percent rate of return, the seventh and eighth at 9 percent, and the ninth at 6 percent. The results of 3, 7, and 9 show how the change in interest rate drives up the cost of the lease plan at $21 per $1,000: If the salvage value is raised from 15 to 30 percent and the rate of return reduced to 9 percent, the cash-purchase plan looks better than the lease plan:

No.	Lease plan,%	Cost
3	12	$23,131
7	9	24,607
9	6	26,181

The SOYD plan is costly and at 12 percent compares unfavorably with the rest. The only thing that would make it attractive would be the high cash flows in the early years if they matched the contractor's earnings during those years. Notice also how

Table 7-12 Comparison of final cost of various plans

No.	Plan	Final cumulative cost
1	Purchase, 12% return	$27,345
2	Bank loan, 6% add-on	23,626
3	Lease, $21 per $1,000	23,131
4	Bank loan, 12% simple	23,871
5	Lease, $22 per $1,000	24,230
6	Lease, SOYD plan at 6% add-on	25,605
7	Lease, $21 per $1,000, 9% rate of return	24,607
8	Purchase, 30% salvage, 9% rate of return	23,539
9	Lease, $21 per $1,000, 6% rate of return	26,181

raising the lease rate from $21 to $22 per $1,000 raises it above the cost of the bank loan.

No.	Plan	Cost
8	Purchase	$23,539
7	Lease, $21 per $1,000	24,607

The moral is that it is very difficult to predict how plans will turn out until they are analyzed in the light of existing circumstances. It is reasonable to say that no one plan of financing is consistently best. When interest rates are known, salvage values can be estimated, and the conditions of the plan are revealed, an economic analysis should be made by the discounted-cash-flow method as demonstrated here. Then, with a knowledge of financial limitations and business objectives, the contractor is better prepared to decide which is the better plan for him at that time.

8
Standardization

GENERAL DESCRIPTION

Standardization is the utilization of like equipment or equipment with identical components. Even though identical makes and models of machines may not be used, a limited degree of standardization can be achieved by the use of the same engine, attachments, or other components in the machines on hand. In discussing standardization, the following definitions will be used:

Type The major category of construction equipment by purpose: crawler tractor, motor grader, wheel tractor-scraper, etc.
Class The size of equipment within a type, usually differentiated by weight, horsepower, cubic-yard capacity, or some engineering measure: the Caterpillar D4 and D7 crawler tractors would be in different classes.
Engine The prime mover in a machine
Engine series Engines of varying horsepower and performance achieved by natural aspiration, turbocharging, aftercooling,

or other design changes and using the same block and number of cylinders

Engine family A group of engine series achieved by using different numbers of cylinders with the same bore and stroke to obtain a great range of horsepowers

The subject of standardization is one which has long been of interest to the users of construction equipment. Forty years ago, the September 1934 issue of *Public Roads*[1] magazine had the following to say:

> The use of a variety of different kinds of equipment has a tendency to increase time losses and decrease production. Equipment is subjected to extremely hard usage and mechanical troubles invariably occur from time to time. It is much cheaper and less difficult to keep an adequate supply of spare parts on hand when the equipment is closely standardized than when a variety of different kinds and sizes of equipment is used. Standardization of hauling units permits interchange of parts and one line of spare parts will suffice for all the hauling equipment. If more than one shovel is employed, there is the same advantage in having them alike. This will permit not only the carrying of a smaller investment, but also operators can be shifted from one piece of equipment to another without impairment of efficiency. Repair men will become more expert in making repairs as well as in diagnosing trouble and in the routine care of the equipment.
>
> Equipment earns no profit except when working. Anything which helps to keep and continue the equipment in working order is therefore of definite value to the contractor. Standardization of equipment so as to permit a wide interchangeability of parts usually requires no outlay and only a little definite planning and fore thought, and should be embraced by all contractors to whatever extent their lines of work will permit.

The economic consequences of equipment standardization have been studied in detail by Koster[2] in his report on the economics of standardization.

---

[1] Power Shovel Operation in Highway Grading, *Public Roads*, vol. 15, no. 7, September 1934.

[2] Francis D. Koster, The Economics of Heavy Construction Equipment Standardization, *Stanford Univ. Constr. Inst. Tech. Rep.* 43, 1964.

LEVELS OF STANDARDIZATION

There are at least three levels of standardization achievable by the contractor:

1. All machines of the same type and class are alike.
2. All equipment is grouped by engine series in several engine families.
3. All machines of the same type and class are alike and all have the same engine family.

In the classification above, the first level is the lowest and usually would be the first step in a standardization program. The third or top level would probably be sought under difficult operating conditions such as a remote location with tenuous supply lines and other difficult problems associated with isolated projects. The amount of standardization to be achieved should be determined by careful planning in advance of the operation. The first or second levels usually are acceptable where parts and labor are readily available, as in most parts of the continental United States.

FIRST LEVEL OF STANDARDIZATION

This is the first step. To choose machines of the same type and class that are alike means that the contractor would select one make and model of each of the following, as needed:[1]

Heavy crawler tractor
    Caterpillar D-8
    Allis-Chambers HD21
    International TD 25
    Terex (GM) 82-40
Medium crawler tractor
    Allis-Chalmers HD 11
    Caterpillar D6
    International TD 15

---

[1] No endorsement of any particular manufacturer is intended. In the lighter classes, there are often many more than appear in the list.

Light crawler tractor
    International TD 9
    Allis-Chalmers HD 6
    Caterpillar D4
    Massey-Ferguson MF 300

These lists may be augmented and other lists prepared for other types of equipment such as motor graders, scrapers, generators, pumps, etc. The most important thing is to keep vintage groups standardized; i.e., all units of any year group should be of the same make and model. If you want to rotate manufacturers of heavy crawler tractors, buy Caterpillar one year, International the next, and so forth. Even better is to continue to buy the same make and model from year to year.

SECOND LEVEL OF STANDARDIZATION

It is possible to improve the first level illustrated above by extending the standardization to various types of equipment using the same engine. Approximately 70 to 80 percent of the spare parts required for support of a machine are for the engine. It then follows that the selection of different machines using the same engine will improve the equipment situation. The Caterpillar D333 engine will be used to illustrate this type of standardization.

The D333 engine is a four-cycle diesel engine with a 4.75-in bore and 6.00-in stroke and has 6 cylinders. The D333NA is naturally aspirated and has a maximum horsepower rating of 175. The D333T is turbocharged and has a maximum horsepower rating of 300. The D334 engine is the same basic engine as the D333 except that it has both turbocharging and jacket-water aftercooling. It is rated at 335 hp.

The following Caterpillar equipment is available with the engines mentioned above:

12F	Motor grader
120	Motor grader
D5	Crawler tractor
561C	Pipelayer
14E	Motor grader
627	Wheel tractor-scraper
980B	Wheel loader
966C	Wheel loader
977L	Crawler loader

STANDARDIZATION

  D6C Crawler tractor
  814 Wheel tractor
  D7   Crawler tractor
      135-kW generator set
      175-kW generator set

Additionally the following is a partial list of original equipment manufacturers (OEM) who use the Cat 333/334 engines in their equipment:

American Hoist & Derrick Co.	60-ton crane
	100-ton crane
	$1\frac{1}{4}$-yd shovel
Harnischfeger Corp.	90-ton crane
	$1\frac{1}{4}$-yd shovel
Insley Mfg. Corp.	$1\frac{1}{2}$-yd backhoe
Koehring Co.	65-ton crane
	$2\frac{3}{4}$-yd backhoe
	$2\frac{3}{4}$-yd scooper
	4-yd scooper
Link-Belt Speeder Co.	1-yd shovel
Northwest Engineering Co.	30-ton crane
	70-ton crane
	90-ton crane
	$1\frac{1}{4}$-yd shovel
	$2\frac{1}{2}$-yd shovel
Skagit Corp.	A-frame yarder
Franklin Equipment Co.	Log skidder
Pettibone-Michigan Corp.	Carry lift
	Log skidder
Barber-Greene Co.	Paver
Construction Machinery Co.	Autograder
R. A. Hanson Co., Inc.	Grader and-finisher
Rex Chainbelt Inc.	Paver
Chicago Pneumatic Tool Co.	Rotary drill
Davey Compressor Co.	Air compressor
Gardner-Denver Co.	Air compressor
Joy Mfg. Co.	Air compressor
Cedarapids-Iowa Manufacturing Co.	Rock crusher
Pioneer Division of Portec, Inc.	Rock crusher
Smith Engineering Works	Rock crusher

THIRD LEVEL OF STANDARDIZATION

The highest level of standardization that can be achieved is when all engines in the fleet are in the same engine family. Here the highest commonality of parts is obtained because all engines have the same size cylinder. Since most of the moving parts in the engine are associated with the cylinder, most of the engine parts in the family will be interchangeable. Three engine families are manufactured by the Detroit Diesel Allison Division (DDAD) of General Motors (GM). These are the 53, 71, and 149 series, all two-cycle diesel engines. They are used in over 3,800 applications by more than 630 original equipment manufacturers (OEM). The 71 series will be used to illustrate engine-family standardization.

The GM 71 series engines all have a bore of 4.25 in and a stroke of 5.00 in. The series number is derived from the fact that each cylinder has a displacement of 71 in^3. These engines were pioneered in 1938 and have been improved continuously with the advent of new technology. The N models have a compression ratio of 18.7:1, while the standard models have 17:1. The family consists of 2-, 3-, 4-, and 6-cylinder in-line engines, and 6-, 8-, 12-, and 16-cylinder V models. Horsepower ratings range from a low of 48 for the 2-cylinder engine to a high of 700 for the 16V-71. Engines are aspirated by both positive-displacement blower and, optionally, turbocharger. Up to 70 percent of all moving parts in the engine family are interchangeable.

Engine	Horsepower
2-71	48–68
3-71	75–106
4-71	101–165
6-71	154–280
6V-71	175–265
8V-71	233–380
12V-71	350–585
16V-71	466–700

Typically, the following equipment can be engined with the 71 series:

STANDARDIZATION

Equipment	Size	Engine
Wheel loaders, yd^3	1½–3	3-71, 4-71
Shovels, cranes, draglines, hoes, yd^3	½	2-71
	¾	3-71
	1	4-71
	1¼–2	6-71
	2–3	8V-71
Wheel-tractor-scrapers, yd^3	7–10	4-71
	15–18	6-71
	18–21	8V-71
	24–40	12V-71
Off-highway trucks, tons	6–12	4-71
	18–20	6-71
	20–30	8V-71
	30–45	12V-71
	45–65	16V-71
	65–85	16V-71T
Air compressors, ft^3/min	250	3-71
	365	4-71
	600	6-71
	900	8V-71
	1200	12V-71
Centrifugal pumps (80–100 lb/in^2) gal/min	Up to 1,000	3-71
	1,000–2,000	4-71
	2,000–3,000	6-71, 6V-71
	3,000–4,000	8V-71
	4,000–5,000	12V-71
	5,000–10,000	16V-71
Crawler tractors, tons	22	6-71
	27	8V-71
	37	two 6-71
Motor graders, hp	108	3-71
	151	4-71
	200	6V-71
Rubber-tired rollers, tons	25	3-71
	35	4-71
Concrete pumps, yd^3/hour	100	6V-71
Generators, kw	20–40	2-71
	40–70	3-71
	60–100	4-71
	100–150	6-71
	125–230	8V-71
	200–340	12V-71
	300–440	16V-71

In addition to the above applications, the 71 series engines are used for many others, both marine and industrial. Because of the importance of engine families in standardization, the families of four major manufacturers of construction equipment will be described.

SOME MAJOR ENGINE FAMILIES IN CONSTRUCTION EQUIPMENT

Four manufacturers supply many of the engines for construction equipment and have engine families which may be used in standardization. They are Cummins, Caterpillar, General Motors, and International Harvester. The Detroit Diesel Allison Division of General Motors has been outstanding in the development of engine families for over 30 years. Very few others have paid much attention to the development of families until recently. Now all seem to be moving in the general direction of developing families and that augurs well for the construction contracting industry.

Cummins has two engine families, the 855/1710 and the new K family. The 855/1710 series covers a range of 250 to 800 hp; 85 percent of all wear parts are interchangeable in the 855 engines and 85 percent of all moving parts with wear surfaces are interchangeable in the 1710 engines. Liners, piston rings, piston pins, fuel pumps, and other miscellaneous parts are interchangeable between the 855 and 1710 engines.

The K family is now under development with the 6- and 12-cylinder engines already tested. Other engines in the family will be introduced in compliance with market conditions and the demands of the industry. In the K family, 90 to 95 percent of the parts will be common and interchangeable. Since the bore and stroke of all K engines will be the same, high parts commonality will exist between in-line and V engines. Figure 8-1 shows the high ratio of interchangeability of parts, using the 6 cylinder

Table 8-1 Cummins engine families

Family	Bore	Stroke	Engine	No. of cylinders	Horsepower range
855/1710	5.50	6.00	855C	6	250–420
	5.50	6.00	1710C	12	500–800
K	6.25	6.25	767	4	70–400
	6.25	6.25	958	5	130–500
	6.25	6.25	1150	6	200–600
	6.25	6.25	1533	8	260–800
	6.25	6.25	2300	12	450–1,200
	6.25	6.25	3066	16	610–1,600

STANDARDIZATION

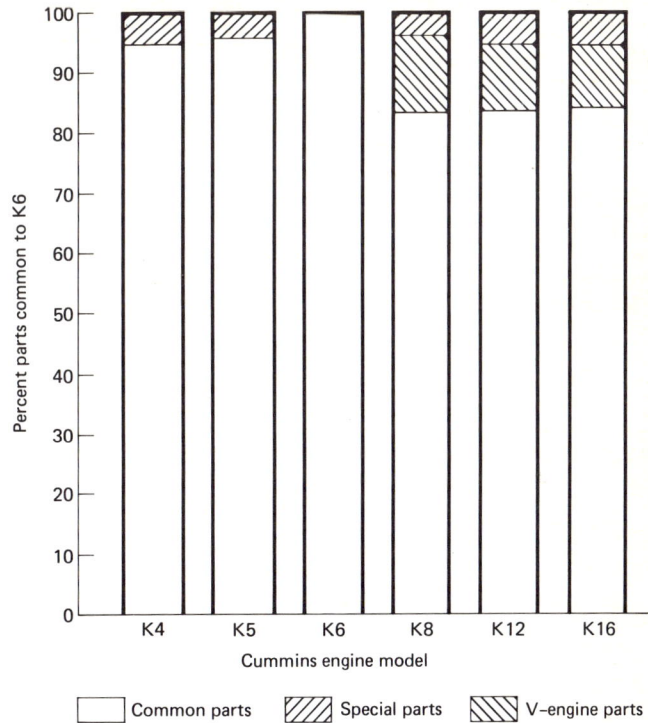

Fig. 8-1 Parts commonality of Cummins K family engines.

engine as the standard.

Table 8-1 shows the Cummins engine families described above. The horsepower range is achieved by varying the revolutions per minute and by adding turbochargers and aftercoolers to the naturally aspirated engines.

There are three Caterpillar engine families that cover a range of horsepowers from 85 to 900. They are the D330, D343, and D353 basic engines with more cylinders, turbochargers, and jacket water aftercoolers to vary the horsepower. The 4- and 6-cylinder engines are in-line, while those with 8 to 16 cylinders are V engines. Bore and stroke remain the same in each family. Table 8-2 shows the Caterpillar families; all are available at the present time (1974).

The 71 series of Detroit Diesel engines has already been used to illustrate the wide application of an engine family. DDAD also manufactures a 53 series and 149 series. All are two-cycle engines, giving them an advantage in weight and size over

Table 8-2 Caterpillar engine families

Family	Bore	Stroke	Engine	No. of cylinders	Continuous bhp range
330	4.75	6.00	D330C	4	85–125
	4.75	6.00	D333C	6	125–190
	4.75	6.00	D334	6	220
	4.75	6.00	1673/1674	6	250–270
343	5.40	6.50	D343	6	245–335
	5.40	6.50	1693	6	325–425
	5.40	6.50	D346	8	445
	5.40	6.50	D348	12	670
	5.40	6.50	D349	16	890
353	6.25	8.00	D353E	6	375
	6.25	8.00	D379B	8	500
	6.25	8.00	D398B	12	750
	6.25	8.00	D399	16	900

the four-cycle diesel engines of equal horsepower. All the Detroit Diesel engines use scavenger air for natural aspiration. The addition of turbochargers and aftercoolers increases the horsepower output. Because the two-cycle engines must have blowers for scavenging, altitude performance is better, and they are generally derated less for altitude than the naturally aspirated four-cycle engines. Bore and stroke remain constant within each family series. Table 8-3 shows the engines in the 53, 71, and 149 series.

Table 8-3 Detroit Diesel Engine Families

Family	Bore	Stroke	Engine	No. of cylinders	Continuous bhp range
53	3.875	4.500	2–53	2	35–47
	3.875	4.500	3–53	3	64–101
	3.875	4.500	4–53	4	87–140
	3.875	4.500	6V–53	6	130–216
	3.875	4.500	8V–53	8	186–283
71	4.250	5.000	2–71	2	48–68
	4.250	5.000	3–71	3	75–106
	4.250	5.000	4–71	4	101–165
	4.250	5.000	6–71	6	154–280
	4.250	5.000	6V–71	6	175–265
	4.250	5.000	8V–71	8	233–380
	4.250	5.000	12V–71	12	350–585
	4.250	5.000	16V–71	16	466–700
149	5.750	5.750	12V–149	12	675–1,000

STANDARDIZATION

International Harvester has two families of engines, the 300 series and the 400 series. Horsepower is varied by use of natural aspiration and turbochargers. In addition, the displacement is varied by changing the stroke and holding the bore constant in each series. About 65 percent of the parts are common to both series, and there is a 95 percent commonality within each series. All engines are 6 cylinders, in-line, with four-cycle diesel design. Table 8-4 lists the 300/400 series engines.

Tables 8-1 to 8-4 illustrate the availability of engine families from some of the prominent manufacturers of prime movers for construction equipment. Other engine families are available. All manufacturers are involved in some standardization in order to enhance production and improve spare-parts support. The best information on this subject is always at the hands of the manufacturers, since engine technology is dynamic and subject to continuous change.

U.S. ARMY CORPS OF ENGINEERS STANDARDIZATION PROGRAM

As a result of World War II experience, the military supply agencies determined that some degree of standardization was required in the military services. The armed forces began studying standardization of engines and accessories shortly after the end of the war. A Department of Defense report[1] placed the blame for many of the problems in military supply squarely on the proliferation of makes and models of mechanical equipment:

---

[1]Department of Defense Report, Engine and Accessories Standardization Program, Jan. 3, 1961.

Table 8-4 International Harvester diesel engine families

Family	Bore	Stroke	Engine	No. of cylinders	Continuous bhp
300	3.875	4.410	D-312	6	80
	3.875	5.085	D-360	6	92
	3.875	5.085	DT-360	6	130
400	4.300	4.750	D-414	6	105
			DT-414	6	150
		5.000	D-436	6	115
			DT-436	6	160
			D-466	6	118
		5.350	DT-466	6	170

The great variety and large amount of equipment used by the Armed Forces created a tremendous maintenance and spare parts problem that required the expenditure of an undue proportion of the total war effort. Approximately fifty percent of the maintenance and spare parts support effort expended by the Military was required to maintain in an operable condition the numerous different internal combustion engines used by the Army, Navy and Air Force. There can be no doubt that the multiplicity of makes and models of engines utilized was the primary contributory cause for the extensive maintenance and supply requirements.[1]

The Corps of Engineers received Army-wide responsibility for studies on the standardization of industrial engines in 1948. Because the results of these studies emphasize the dramatic savings which can be achieved by a modest amount of standardization, they are interesting to users of equipment, both military and civilian.

In August 1948 an Industry Advisory Committee for Industrial Engines was established to advise the Army on a program. Membership on the committee consisted not only of military supply representatives but also representatives from the major engine manufacturers. A four-part program was proposed:

1. A standardization of high-mortality parts for industrial gasoline engines
2. A diesel-engine program
3. Development of a family of small gasoline engines
4. A standardization of engine accessories

The Advisory Committee guided the work on the program and reviewed the results.

In the standardization of high-mortality parts for gasoline engines the number of such parts was reduced from 2,872 to 199. Figure 8-2 gives a breakdown of this impressive reduction in wear parts. This program covered 12 basic bore sizes of engines ranging from $2\frac{1}{4}$ through $5\frac{3}{4}$ in and a horsepower range of approximately 10 to 300. By standardizing the wear parts such as valves, bearings, pistons, etc., a 93 percent reduction of parts was effected.

[1] Ibid., p. 3.

STANDARDIZATION

Item	Number of parts	
	Military standard	Commercial
Valves	32	383
Connecting rod bearings	24	187
Plain main bearings	40	509
Flanged main bearings	25	153
Connecting rod bearings	9	194
Pistons	19	203
Piston pins	11	194
Piston rings	39	1,045
Total	199	2,868

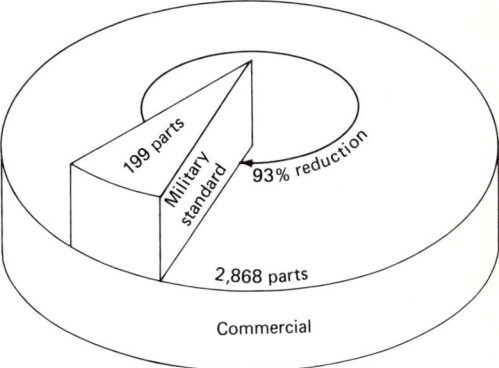

Fig. 8-2 Reduction from standardization of high-mortality parts.

The committee decided that a standardization of diesel engines by means of interchangeable parts was impractical. To accomplish a limited standardization in this area, a program of qualification testing was established to qualify specific engines and standardize on them for procurement.

The need for development of a small family of air-cooled gasoline engines was brought out by studies of engine procurement during World War II. Figure 8-3 shows the number of industrial engines procured versus the engine-displacement range. It is impressive that 73 percent of all gasoline engines procured were less than 20 hp and that there were 78 different makes and models.

The family of small industrial engines consisted of six models ranging from $\frac{1}{2}$ to 20 hp. Table 8-5 shows the characteristics of

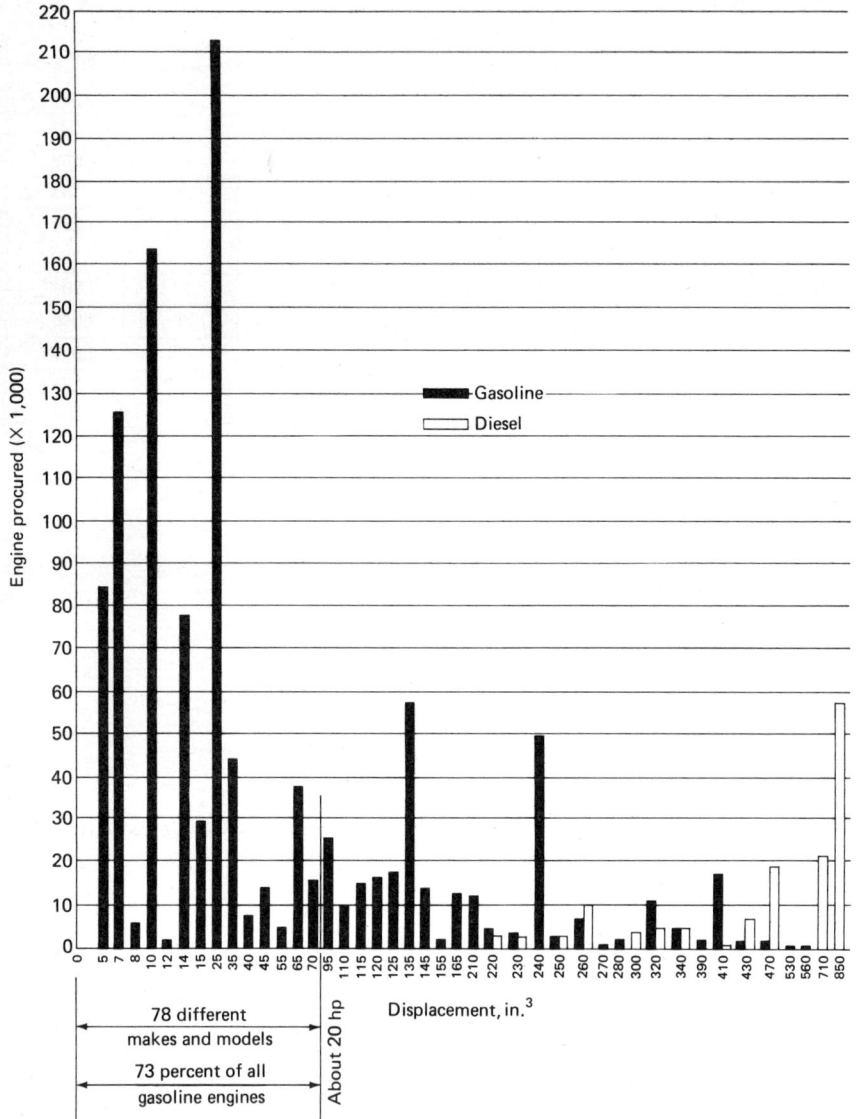

Fig. 8-3 Industrial engine procurement, all services.

this group with engines of three different bores and strokes. They all are four-cycle, gasoline, air-cooled engines with overhead valves, varying from 1 to 4 cylinders.

STANDARDIZATION

Table 8-5 Military family of small industrial-type gasoline engines

Family	Bore	Stroke	Engine	No. of cylinders	Continuous rated hp
1.75	1.75	1.25	1A03	1	0.50
2.25	2.25	2.00	1A08	1	1.50
	2.25	2.00	2A016	2	3.00
	2.25	2.00	4A032	4	6.00
3.00	3.00	3.00	2A042	2	10.00
	3.00	3.00	4A084	4	20.00

It is estimated that approximately 23,000 different parts were required to support the 78 makes and models of these small engines procured in World War II. Approximately 700 different parts will be required to support the family shown in Table 8-5. Actually, it is planned to stock only 40 different line items, consisting of subassemblies and repair kits, to support the six engines in the family.

In the accessory-standardization program, equally beneficial results were achieved. The accessories considered for standardization and the results achieved are shown in Fig. 8-4. In this case, total accessories and spare parts for them were reduced from 16,156 to 210.

From these Corps of Engineers programs, it is abundantly clear that some standardization can be useful to the civilian contracting industry as well as the military. One has only to look at the proliferation of makes and models in the average storage yard to realize that some thought should be given to minimizing the different makes on hand.

AN EXAMPLE OF THE ECONOMIC BENEFITS

Shortly after the beginning of the Korean War, Navy Seabees were sent to Subic Bay in the Philippines to build a major base. Since this was basically an earth-moving job, a large amount of construction equipment was shipped from available sources in 1951 to get the job started. Over the next couple of years, more available equipment was shipped in to speed up the job. By the end of 1953, the equipment situation had become serious because of the lack of standardization.

In order to get a grasp of the situation, the Bureau of Yards and Docks (now the Naval Facilities Engineering Command) awarded a contract to the Bechtel Corporation of San Francisco to study and report on the equipment situation in the Philip-

Fig. 8-4 Comparison of total repair-parts support for engine accessories. (Data from U.S. Army Corps of Engineers.)

Accessory	Military standards			Commercial		
	Number of standard accessories	Number of repair parts required (each accessory)	Total number of repair parts required	Number of commercial accessories	Number of repair parts required (each accessory)	Total number of repair parts required
Air Cleaners	8	0	8	280	6	1,680
Regulators	3	0	3	262	5	1,310
Fuel Filters	3	0	3	75	2	150
Fuel Pumps	3	0	3	280	4	1,120
Generators	3	13	39	262	6	1,572
Governors	4	3	12	295	6	1,770
Magnetos	6	2	12	252	14	3,528
Mufflers	8	2	16	225	2	450
Oil-Filter elements	6	0	6	35	0	35
Radiators	10	0	10	125	0	125
Starting motors	6	4	24	293	6	1,758
Thermostats	7	0	7	137	0	137
Total	67	24	143	2,521	51	13,635
Total accessories plus repair parts			210			16,156

STANDARDIZATION

Fig. 8-5 Reduction by family standardization.

pines. Figure 8-5 shows the results of the Bechtel report[1] on reduction of equipment, manpower, and costs which would result from a limited standardization on equipment families. Figure 8-6 shows the comparison of conditions before and after this standardization. The report sums up these benefits by saying:

> The economic benefits derived when the recommendations are implemented will result in a 24 percent reduction in

---
[1] Bechtel Report on Contract NOy 75562, U.S. Navy, July 1954.

Fig. 8-6 Comparison of conditions before and after standardization.

Item	Family standardization		
	Before	After	Benefits
Total units of equipment	1,462	1,111	351 reduction
No. of manufacturers, major equipment	110	77	33 reduction
No. of models, major equipment	232	133	99 reduction
Daily average of earthmoving, yd^3	23,414	31,920	8,506 increase
Daily average of aggregate production, yd^3	600	2,400	1,800 increase

equipment, a 25 percent reduction in manpower and a decrease in the investment of equipment and spare parts of approximately $4,535,459.... An analysis of recommendations indicates a considerable increase in production and lowering of the production costs for materials and supplies. These economic benefits have been achieved by utilizing the principle of standardization in adapting a "Family of Equipment" for the entire Subic Bay area program.[1]

This is an unusual case—lots of equipment, high-priority work, remote overseas location, all worsening the effects of having many makes and models of equipment. Even though the number of makes was reduced only from 110 to 77 and the number of models from 232 to 133, the savings in parts and equipment alone were considerable. Just think what could have been accomplished if engines families could have been standardized before the job started.

SOME BENEFITS OF STANDARDIZATION

There are a number of economic benefits of standardization, not all of which were mentioned in the example above. Some of these benefits are intangible and thus difficult to reduce to a dollar value.

Probably the easiest benefit to discern is the saving in spare-parts inventory due to commonality of parts. The matter of parts interchangeability has already been mentioned in the discussion of engine families. When more pieces of like equipment are grouped together, the cost of spare-parts inventory is reduced. The size of the inventory of parts is affected by a number of factors:

---

[1] Ibid., Equipment Survey, p. 1.

1. Isolation of the job, which increases procurement time
2. Number of machines per engine family
3. Severity of service
4. Penalty costs of downtime
5. Average age of equipment
6. Adequacy of spare-parts list

Economically the contractor must seek to balance the cost of downtime against the cost of spare parts in the inventory. Where downtime costs are high and procurement time is great, more parts must be carried in stock to avoid deadlining equipment. In domestic areas where parts are readily available from a local supplier in a matter of hours, the contractor's inventory will consist mainly of consumables and fast-moving items. In remote areas where lines of communication are extended, the contractor must carry insurance items in stock as well as the consumables. Consumables for 2,000-hour support of equipment will run from 2.5 to 3.5 percent of the original cost of the machine; the remaining costs are for insurance items. Curves showing the recommended level of parts support for heavy-construction equipment in overseas service were developed by Koster.[1] Figure 8-7 shows the parts support for easy, average, and severe service overseas. Note how rapidly the inventory costs drop with an increase in the number of like units (standardization) and how the type of service affects the inventory requirement. The cost of inventory maintenance will be 15 to 25 percent of the average inventory level per year; the latter figure applies to a remote construction project. This charge includes labor for handling and accounting, insurance, warehouse rental, transportation, depreciation, damage and loss, and interest on the investment.

A second benefit from standardization will be the lower cost of maintenance, because mechanics are more efficient when working on like machines. Because there are fewer kinds of machines, less training will be required to develop efficient mechanics with a knowledge of repair procedures.

The use of learning curves to develop the time to accomplish a certain repair is discussed at length by Koster.[2] Mechanics working on a single engine model are able to accomplish a given repair in less time than an equal mechanic who works on several

[1] Op. cit., p. 23.
[2] Op. cit., chap. 3.

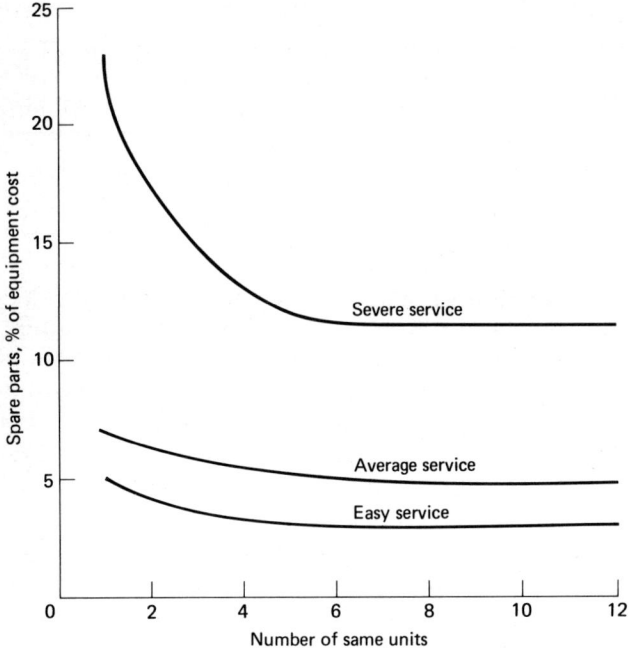

Fig. 8-7 Parts support for overseas service. *(After Francis D. Koster, Stanford Univ. Constr. Inst. Tech. Rep. 43, 1964)*

different engines. Table 8-6 summarizes the opinions of various equipment managers on the reduced repair time required because of familiarity.

Besides lower maintenance-labor costs, lower operating-labor costs are another benefit of standardization. If machines are alike, it is easy to shift operators in case of illness or indisposition of a regular operator. Less operator training is required as a result of standardization. Operators are generally more

Table 8-6 Reduced repair time resulting from engine familiarity

Engine category	Percent of standard hour
Single model (4 or more units)	80
Single family (4 or more in family)	85
2 different models (4 or more in 2 models)	90
2 different families (4 or more in 2 families)	95
Nonstandard engines (4 or more)	100

proficient because they are familiar with the controls and handling characteristics of like machines. Safety, too, is enhanced because of increased operator proficiency. In an emergency, an operator is more likely to grab the correct control if he operates only one model of equipment. These costs are difficult to reckon, but like all intangibles, it is better to recognize them at some value than to discount them as zero. The owner's evaluation of these costs will have to be subjective except as they may reduce his insurance costs through better safety.

In standardizing on certain equipment, the owner generally enjoys a better relationship with his dealer. Having more leverage with the dealer always pays off in times of short supply. And by having closer ties with the dealer, he can expect to get better dealer service.

Finally, the administration of a mixed bag of equipment is more expensive and less efficient because of increased paperwork. There are simply more people involved in getting parts support, in getting advice, and in liaison with the manufacturer than when only a few makes of equipment are involved.

SOME DISADVANTAGES OF STANDARDIZATION

Not everything related to standardization is to the advantage of the contractor. For one thing, competitive bidding is out unless you have a large enough spread to enable you to accept two families of equipment (or engines) instead of one. If you go out on bid, the very manufacturer on whom you standardized may not respond. Even worse, he may up his price figuring that you are trapped. There is at least one instance on record where a governmental agency purchased 20-ton crawler tractors on a competitive bid while another branch of the same agency had to pay an additional $2,000 each to purchase by negotiation 15-ton crawler tractors on which they had standardized. Much of the final price on a negotiated purchase under a standardization procedure will depend on how much leverage the buyer can bring to bear.

Another disadvantage of standardization is that it is difficult to reject a dealer without breaking the adopted standard. Since most dealers are franchised by the manufacturer to sell in a given geographic area, it is difficult and sometimes impossible for the buyer to go to another area to buy the same product. If a contractor is large enough, he can do this, but he then accrues additional shipping costs.

A third disadvantage is that if there is a weakness or bug in one machine, it is very likely that all the others have the same weakness. If this happens to be goosenecks on scrapers, first one breaks and soon all are deadlined. Any other inherent weakness can do the same thing. Only when a model has been in production for several years can one assume that all the bugs are known and designed out of it. Never has a completely new machine been introduced to the market without some weaknesses. Great caution should be exercised in adopting a brand-new design for a standard. It is almost always better, in fact, to try out a single machine of this type for at least a year before deciding to standardize on it.

Probably the greatest economic disadvantage in standardization is the sometimes additional cost of the OEM installing an optional engine. Certain makes of engines command a higher price on the market. To have these engines installed in your equipment may add as much as several thousand dollars to the cost of each machine. This should be investigated carefully before any standardization is instituted. It is also true that the machine may be worth several thousand dollars more on the secondhand market because it has this particular engine in it. At any rate, these costs should be considered in selecting an engine family for standardization.

EXAMPLE OF THREE LEVELS OF STANDARDIZATION

The best way to illustrate standardization is by an example of how a contractor might select equipment to fulfill the requirements of the three levels defined at the beginning of the chapter. Assume that a contractor has a large earth-moving job of approximately 3 years duration. Table 8-7 is a partial list of his major equipment, which he plans to purchase for the work. The total cost of this equipment FOB plant is about $4.5 million. Three levels of standardization are shown in the three columns listing the makes and models of equipment. The worst condition of having a completely heterogeneous group of machines is not shown. This would be a situation where the contractor used several different makes, and perhaps different sizes, of scrapers and crawler tractors.

Level 1 standardization is shown in the first column listing the equipment. Here is standardization within types and class. All twin-engine scrapers of the same class (size) are alike, e.g., Caterpillar 657 and Caterpillar 627; all single-engine scrapers are Terex S-32; all 25-ton crawler tractors are International TD25C; and all 35-ton tractors are Caterpillar D9G. This is about the minimum standardization any contractor should accept unless

Table 8-7 Example of three levels of standardization

Item no.	Quantity	Description	1 No family		2 Caterpillar family		3 Detroit diesel family	
			Equip.	Eng.	Equip.	Eng.	Equip.	Eng.
1	4	Scraper, 32 yd³	CAT 657	D346 D343	657 651	D346 D343	TS-32 S-32	12V-71T 8V-71T
2	16	Scraper, 32 yd³	TEREX S-32	DD 12V-71T		D346		12V-71T
3	2	Crawler tractors, 25-ton	IHC TD25C	IHC DT1817B	D8H	D342	82-40	8V-71N
4	5	Crawler tractors, 35-ton	CAT D9G	CAT D353	D9G	D353	82-80	2/6-71N
5	2	Motor graders, 15-ton	WABCO 777B	CUM N743C	14E	D333	AW Super 500	6V-71
6	2	Scrapers, 14 yd³	CAT 627	CAT D333	627	D333	TS14B	4-71N 4-71N
7	2	Pneumatic rollers	BROS SP 10000	CUM JN-6	BROS SP 10000	D330	BROS SP 10000	4-71
8	1	Tamping foot roller	BROS SP446S	DD 8V-71N	825B	D343	BROS SP446S	8V-71N
9	1	Generator, 150-kW	ONAN	CUM NT-855P	CAT	D334	GM	8V-71
10	1	Compressor, 900 ft³/min	IR	DD 8V-71N	GD	D343T	CP	8V-71N
11	1	Truck crane, 25-ton	AM 375BT	IHC/RD501 WAU/135GZU	AM 375BT	1673 D330T	AM 375BT	4-71 4-71
12	1	Gradall (G-800)		IHC/RD501 DD 4-71		1673 D330		4-71 4-71

Number of engine families

| | | | | 10 | | 4 | | 1 |

Key:
AM	= American Hoist	CP	= Chicago Pneumatic	GM	= General Motors	ONAN	= Onan Electric
AW	= Austin-Western	CUM	= Cummins Diesel	IHC	= International Harvester	TEREX	= Terex Div., GM
BROS	= Bros	DD	= Detroit Diesel, GM	IR	= Ingersoll-Rand	WABCO	= Westinghouse Air Brake
		GD	= Gardner-Denver			WAU	= Waukesha

there are compelling and substantial reasons for not doing so. In level 1 there are 10 different engine families. Without any effort to standardize there could have been at least 20 families, since there are 36 machines. So with this level of standardization, the number of engine families has been at least halved.

In level 2 standardization, the contractor has adopted a Caterpillar line of machines. By doing so the number of engine families has been reduced to four:

1. D330/D333/D334/1673
2. D342
3. D343/D346
4. D353

Now if the contractor should decide that he could substitute either D7 or D9 crawler tractors for the D8s, he could eliminate the D342 engines and get down to three engine families instead of four.

In level 3 standardization, the contractor opted for a complete line of equipment with Detroit Diesel's 71 series engines. In this case, there is only one engine family. Here there is a maximum interchangeability of parts because of the common cylinder size. While at present the Detroit Diesel family is the only one that has such a broad range of application, the new Cummins K family should be capable of doing the same. Caterpillar could take the 4.75 × 6.00 bore and stroke used in the D330/D333/D334/1673 series and design a family of 3-, 4-, 6-, 8-, and 12-cylinder engines which could meet the requirement of a single engine family. And why not?

CONCLUSIONS ABOUT STANDARDIZATION

Some degree of standardization will always be beneficial. How far a contractor should go will depend on his evaluation of the economic benefits to be gained. Some of the benefits and costs which can be directly calculated are:

1. Savings in maintenance labor
2. Savings in parts inventory
3. Savings in downtime
4. Cost of installing optional engine

Other advantages and disadvantages are intangible and will have to be evaluated subjectively according to the situation at the time. Some of these factors are:

1. Effect of dealer monopoly on service
2. Administrative savings
3. Savings in training costs
4. Leverage in purchasing equipment
5. Savings in operating labor

The calculable economic benefits of standardization have been estimated by Koster to range normally between 5 and 15 percent of the original cost of equipment.[1] In special cases of remote operations where the lines of support are tenuous, the benefits may be several times this 15 percent maximum. A contractor should invariably try to achieve a first level of standardization as minimum policy. The time to plan standardization is before the job starts rather than afterward. Where equipment is used on a continuing basis, standardization policy can be implemented by phases with due regard to requirements of the business and budget cycles. Second-level standardization can be a logical step up from the first level. It should be easily achievable with foresight. The third level of standardization is more difficult and will require more careful analysis to implement. When more manufacturers come out with good engine families, the third level will be more easily achievable. Meanwhile, the contractor should never settle for less than a first-level standardization.

[1] Op. cit., p. 76.

9
Inventory Management

DEFINITION

One of the most neglected aspects of equipment policy is management of the equipment inventory. By inventory management is meant registration (establishment of a numbering system), keeping inventory records and making periodic inventories of all eligible equipment, maintaining proper storage of the equipment, guaranteeing security, and assigning custody. It is unfortunate that many owners consider these items of insufficient importance to merit top-level attention. The result is a gradual but sustained loss of value and materials which could be saved.

REGISTRATION SYSTEM

The first step required in developing a responsible inventory of equipment is to adopt a registration system by which permanent contractor numbers can be assigned to each machine as it is acquired. A system of this sort not only serves to identify each machine but also aids in recording the history of the machine as it passes through its working life, and in collecting costs and operating statistics which are so important in analyzing its performance. A logical numbering system can best be developed by referral to an equipment classification code. Four such classification codes are the Swedish system,[1] APWA code,[2] the Nav Fac Equipment code,[3] and the FHWA code.[4] The code which is proposed herein is similar to these but more slanted toward construction equipment with less emphasis on equipment not often used in construction contracting.

At the time of inspection on receipt, a permanent contractor identification number should be given to the piece of equipment. This number should be permanently affixed to the machine and its records and not changed during possession by the purchaser. Changing equipment numbers or having two machines at the same time with the same number creates a very difficult situation with regard to machine identification. Registration numbers should be permanently stamped on the manufacturer's serial-number plate if there is room, or a new plate with the number on it should be attached to the machine. In addition, these same numbers should be painted on the machine in a location and manner which will be described later.

When the registration system is established, the eligibility of equipment which will receive these numbers should first be determined. A heavy-construction contractor will usually find that about 75 percent of his capital investment in equipment is in about 25 percent of his machines. These are the first and most important machines to get into the system. Thereafter, some dollar limit should be set on the minimum value of a machine to be registered. A good minimum to start with is $500: any machine which costs more than this amount should be given a number for permanent contractor identification. If this does not strain the accounting system which is invoked with registration, the dollar value might be lowered to $100. This is a matter of judgment.

---

[1] Issued by the Associated General Contractors and House Builders of Sweden (Svenska Byggnadsentreprenörföreningen) in 1968.
[2] American Public Works Association, *Motor Vehicle Fleet Management*, August 1970.
[3] Management of Transportation Equipment, *Man. Nav. Fac.* P-300, Naval Facilities Engineering Command, September 1971.
[4] Published by the Federal Highway Administration (formerly Bureau of Public Roads), February 1971.

INVENTORY DATA CARD

An inventory data card must be filled out for each eligible machine on the inspection when it is received. At this time it will receive its permanent registration number in accordance with the numbering system described in the next section. Figure 9-1 is a sample inventory data card. It can be modified to fit best the fleet which is under consideration. Important aspects of the entries are discussed by box number below. The purpose of keeping the information should be made clear so as to teach the record keeper the necessity of accurate data.

Fig. 9-1 Sample inventory data card.

INVENTORY DATA CARD						1. Equipment number		
2. Description (see Equipment Classification Code)								
3. Equipment manufacturer		4. Yr. of mfr.	5. Model number			6. Serial number		
7. Engine manufacturer		8. Yr. of mfr.	9. Engine model		10. Horsepower	11. Engine serial number		
ACCESSORIES	12. Description			13. Mfr.		14. Yr. of mfr.	15. Model no.	16. Cost (fob)
17. Tires: axle	Front 1	2	3	4	Rear 5	22. End of Yr	23. Mo/Yr	24. Mkt. Value
18. Number						0		
19. Type						1		
20. Size & ply						2		
21. Cost (fob)						3		
25. P.O. number	26. Seller	27. Delivery date	28. Disposal date	29. Life (months) Estimated / Actual		4		
COST SUMMARY	30. Cost (fob)	31. Freight	32. Taxes	33. Total cost	34. Weight (pounds)	5		
35. Basic machine						6		
36. Accessories						39. Selling price		
37. Tires						40. Cost of disposal		
38. Total						41. Net salvage value		

INVENTORY MANAGEMENT

Box 1 is the assigned equipment number, which is next sequential number in the record book in the same equipment-classification code, discussed with the development of a numbering system. Box 2 is the description of the equipment as found in the classification code. Box 3 is the name of the original equipment manufacturer, 4 is the year of manufacture of the basic machine, while 5 and 6 are the model number and serial number. Box 7 is for the name of the manufacturer of the engine. Boxes 8, 9, 10, and 11 relate to the engine. If there are two engines, as in a twin-engine wheel tractor-scraper or a siamese tractor, both should be listed here with the front engine at the left followed by a slash and then the rear engine. For example, the Caterpillar 657 scraper would show as engine models D346/D343.

Accessories and attachments are described in boxes 12 through 16. Box 12 is a short description of the item. Boxes 13, 14, and 15 are the name of the manufacturer, year of manufacture, and model number of each of the accessories listed under box 12. Box 16 is the FOB cost of the particular item. All major accessory items should be listed here including radio equipment, power control units, dozer blades, and other optional equipment other than special tires included in the purchase price.

Boxes 17 through 21 are information on tires for wheeled equipment. Box 17 lists the axles from front to rear with spaces for one to five axles. Boxes 18, 19, and 20 show the number of tires (wheels) and the size and ply rating. If you are using ton-miles-per-hour ratings, they can be shown in box 19 with the type of tire. Box 21 shows the total cost of tires on a particular axle.

Boxes 22, 23, and 24 are to be used for recording annually the market value of the machine in order to obtain a capital-cost curve for analytical purposes. Box 22 is the end of year starting at 0, the date of receipt of the equipment. Subsequently, at the end of each year, the market value of the machine and accessories should be estimated by the equipment manager or his representative. Month and year of the appraisal should be recorded in box 23 as 6/73, meaning June 1973. In box 24, the market value should be recorded.

Box 25 records the purchase-order number, box 26 the name of the seller, and box 27 the delivery date of the machine when purchased. Box 28 records the date of disposal and box 29 the life in months. The estimated life is recorded when the machine is purchased and put into service. The actual life is recorded on the date of disposal by subtracting box 27 from box 28.

Boxes 30 through 41 are a summary of the capital costs of the machine, accessories, and tires. The acquisition costs are broken down into FOB cost, freight, and taxes. The weights are useful in figuring freight and shipping costs and may be useful later to identify the particular model of the machine. The cost profile ends with box 41, which is the net salvage value. All costs should be recorded in dollars of that year. In other words, no attempt should be made to convert them to constant dollars. That can be done by the analysts who use them, since the dates of the costs are also recorded.

REGISTRATION NUMBERING SYSTEM

Equipment numbers consist of five or six digits starting with the three numbers of the equipment classification code and followed by two or three digits to designate the machine. The equipment classification code breaks the equipment down into groups, types, and classes, each being a subdivision of the next higher order. The inventory is broken down by groups as follows:

Group	Description
0	Passenger-carrying and emergency vehicles
1	Highway trucks and trailers
2	Aggregate production and paving equipment
3	Drilling, pile-driving, and compaction equipment
4	Excavating and loading equipment
5	Grading and hauling equipment
6	Miscellaneous construction and maintenance equipment
7	Railway, mining, and marine equipment
8	Weight-handling equipment
9	Materials-handling equipment

These groups are further subdivided into types. Appendix 5 is a list of the groups broken down into type codes. The type codes should be broken down further by the user to reflect the classes of equipment in his inventory. For example, type code 52 crawler tractors might be broken down into class codes as follows (see Table 9-1).

The last two numbers (or three, if required) are the numbers of the particular machine. This series starts with 01, 02, 03,... and ends with..., 98, 99, 00. With two digits, 100 machines can be numbered. If the fleet is very large, three machine numbers can be used: 001, 002, 003,..., 998, 999, 000. This will take care of 1,000 machines of the same class.

INVENTORY MANAGEMENT

The complete registration number consists of the group code, type code, class code, and machine number. With the class codes in Table 9-1, the registration number of a Caterpillar D-9 tractor might be 52035 as follows:

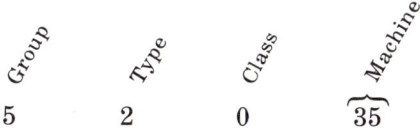

LOCATION AND SIZE OF NUMBERS

Equipment registration numbers should be painted on each machine in numerals 3 or 8 in high in a color which contrasts with the basic color of the machine. They should be easily visible while the machine is working or in motion. The number may be preceded by the initials or insignia of the contractor in a design of appropriate size.

The 3-in numbers should be used on most equipment. The 8-in numerals will be useful on large moving equipment where load counts are made or the identity of the machine is required while it is in motion. Large front-end loaders, dump trucks, wheel tractor-scrapers, and the like will be candidates for the 8-in numbers. An example of the numerals is shown in Fig. 9-2.

These numbers are best located on the engine compartment or cab door of a machine. Where useful they may also be located on the front or rear of a vehicle. The 8-in numbers should be sized of 1-in-width lines instead of $\frac{1}{2}$-in-width lines used in the 3-in numbers.

Table 9-1 Example of extension of classification code to class codes

Group code	Type code	Class code	Description
5	2	0	Crawler tractor, over 350 hp, diesel engine
		1	Crawler tractor, 250–350 hp, diesel engine
		2	Crawler tractor, 140–250 hp, diesel engine
		3	Crawler tractor, 70–140 hp, diesel engine
		4	Crawler tractor, less than 70 hp, diesel engine
		5	Crawler tractor, gasoline engine
		6	Crawler tractor, pipelayer
		7	Other crawler tractors
		8	Vacant
		9	Vacant

   **52035**

Insignia   Registration number

Fig. 9-2 Example of equipment number.

Where equipment is operated during the hours of darkness, reflectorized paint can be used for the registration numbers to augment other safety markings and enhance the visibility of the number. Decals, reflectorized or not, may be used for both insignia and numerals. Federal Specification L-S-300 may be useful in specifying reflectorized pressure-sensitive film for decals.

STORAGE

It often becomes necessary for the contractor to store his equipment for various lengths of time. When a machine is idle for only a few days, no special attention is required unless weather conditions are extremely adverse. Under reasonably temperate conditions a machine may be idle for several weeks without any special care or precautions other than those necessary to prevent weather damage (rain, freezing, etc.). If a machine is to be idle for more than a couple of weeks, it should be prepared for short-term storage. If it is to be idle for more than several months, it should be prepared for long-term storage. The degree of preparation and care required for either short- or long-term storage will depend on the severity of the climate.

The military services have had considerable experience with storage of equipment and should be consulted when the equipment manufacturer is not able to supply adequate instructions. Manuals have been written on how to prepare and preserve equipment for both short-term (live) storage and long-term (dead) storage. Storage conditions will vary considerably depending on the location and existing conditions. All circumstances affecting the stored equipment must be considered in determining what degree of preparation and how much preservation are required. Hot, humid conditions are the worst, especially near salt water. Storage under cover of a roof is far better than outside, especially where rainfall is heavy.

INVENTORY MANAGEMENT

SHORT-TERM STORAGE

Short-term storage is generally necessary between jobs and usually will be out in an open storage yard. Equipment should be started up and exercised at least once a week to keep parts lubricated and the battery charged. Exhaust pipes must be kept covered to prevent the entry of moisture, either rain or dew. This should always be done even when equipment is only parked overnight. Tires should be kept properly inflated and not be permitted to stand in the same footprint (weekly exercising usually will take care of this).

Care should be taken to see that cabs are closed and not invaded by birds or insects. Fluid levels should be checked before equipment is exercised. If oil pressure does not build up to normal 30 seconds after the engine is started, the engine should be stopped and the cause of improper operation determined. All equipment in short-term storage should be inspected at least weekly to check for abnormal deterioration, petty pilferage, and acts of vandalism. Fuel tanks should be kept filled and locked. This will prevent condensation of moisture in the tank, which will cause rusting and improper engine operation. A locked tank will discourage petty theft of fuel and vandalism.

The location of the storage yard is important for the proper care of the equipment as well as its security. Preferably this should be in a compound near the contractor's repair shops. This makes it convenient to properly store and maintain surveillance of equipment awaiting repairs. The storage yard should be well drained and have a gravel or rock surface without potholes or mud baths. Prevailing winds should not blow dust over the equipment from contiguous fields or dusty access roads. The yard should be nearby to provide easy access to fuel, lubricants, water, and other necessary consumables.

LONG-TERM STORAGE

If a machine is to be stored for more than several months, special steps are necessary to see that it is properly preserved and protected over the interim. A dirty machine should never be placed in storage for any length of time. It should first be thoroughly cleaned of all dirt, mud, debris, or anything else that might serve as a protective cover for moisture. Follow the manufacturer's instructions as far as possible. Storage under cover is best. Batteries and tires can be removed and stored best in refrigerated storage, if available and economical. If it is not available, batteries can be placed on a trickle charge and tires stored in a dark, cool place. Seals, insulation on wires, gaskets, and other rubber and fabric parts will deteriorate with age, especially under adverse climatic conditions. They should always be checked carefully before the equipment is put back in operation.

If tires cannot be removed, it is wise to block up the axles on the vehicle to prevent a permanent set in the tread and sidewalls. After 4 or 5 years of outside storage, the tires should be replaced anyhow before the equipment is placed in operation.

For long-term storage it is especially important to seal up all entrances to the internal parts of the engine. Exhaust lines should be plugged and so should the air intake. Whether or not the engine and other metallic parts are coated with Cosmoline or other preservative material will depend on the circumstances. This kind of preservation and consequent depreservation is time-consuming and costly. Before it is entertained as an alternative, serious consideration should be given to whether the equipment should be retained at all.

INFLUENCE OF CLIMATE ON PRESERVATION

Some climates are so severe that even a few weeks idleness without proper protection will result in damage to an engine or its components. This is particularly true of tropical climates, especially near the ocean. If an equipment storage yard is near the sea, salt-laden air will enter the engine through open valves and rust or corrosion will start quickly. It is best to locate equipment storage yards at least $\frac{1}{2}$ mi back from the beach.

A desert climate is destructive to rubber and fabric parts but not to metallic parts. The high temperature and low humidity generally prevent the condensation of moisture on or in the machine, and hence the hot, dry climate enhances preservation of metal. Since sand is usually present in this kind of climate, steps should be taken to prevent the entry of sand into the engine and bearings or other wear parts associated with friction.

Extreme cold is also a good preservation atmosphere for equipment. Since there is very little moisture in air at temperatures below $-20°F$, there is negligible deterioration in metals stored below this temperature. Additionally, rubber and synthetic materials are well preserved at low temperatures and suffer little damage with aging. Wherever equipment is stored in areas subject to very cold winters, care must be taken either to drain the water from the cooling system or to replace it with enough antifreeze to protect it from freezing. One must be especially warned against shipping liquid-cooled engines by ships that may travel a great-circle route through the Arctic at a time

INVENTORY MANAGEMENT

when the shipping point and destination are both in warm climates.

Climate can be either a help or a hindrance in storing equipment. In the last analysis, the owner must make a very careful assessment of the costs and benefits of storage and retention versus disposal of the equipment and replacement at a later date when required. There is a great tendency to retain an old machine on the premise that it is fully depreciated without recognizing the share of storage cost, insurance, security, supervision, and other overhead charges which the machine should justly bear.

SECURITY

As the inner core of the big city has grown, ghettos have formed and crime has increased. The sociological consequences of our inversion of population from rural to metropolitan have really begun to press upon us in the past decade. Any contractor working equipment in a metropolitan area has learned the consequences of relaxed security in terms of dollars. Vandalism, mischief, and petty theft have become costs which must be reckoned in the cost of doing work in the city, whether on highways, streets, overpasses, or buildings.

VANDALISM

Most acts of vandalism can be charged to three groups of people: children who are committing malicious mischief, misdirected social activists who see themselves as saviors of the environment or wish to focus attention on some cause, and union workers who are trying to destroy a nonunion contractor. Since the dedication and expertise of these three groups vary, the contractor must use a mixed bag of tricks to protect himself from each.

Construction equipment has always been an attractive nuisance to children. The contractor's largest exposure to this type of vandalism occurs when the job is located near a school, a teen-age hangout, or the inner city. If school children pass a construction site after the job has been secured for the day, they will naturally want to play on or with the machines parked on the site. To diminish the danger from this type of activity, machine controls should be locked, fuel-tank caps should be locked, dozer blades and booms should be on the ground. Some contractors have found that education of the children at a nearby school also helps prevent dangerous mischief. Sound movie films on equipment safety can be obtained and shown at the school with the cooperation of school officials. Danger-warning signs and "no trespassing" signs should also be placed on the construction site. Parents in the community can also be warned about the dangers of playing on construction sites through the news media and by contractor participation in the local school parent-teachers organization and local service clubs such as Lions, Rotary, Optimist, etc.

Teenage hangouts are also a proximate cause of juvenile vandalism. Parks and playgrounds and even ice-cream parlors can be the focus of adolescent activity. It is more difficult to combat vandalism here than near a school, for the children cannot be reached through a single organization such as a school. Signs, locks, and perhaps a guard may be necessary for these locations. Teen-agers are more malicious than younger children of early school years, who are more likely to play on the equipment and get hurt. Teen-agers pass through a stage where they like to destroy for the sheer thrill of destruction. This will generally be worse and more costly than the damage caused by the young mischief-maker.

In the inner city, children of all ages are apt to congregate at the construction site to play on weekends, especially if the job is not working and the equipment is unattended. This is particularly true where there are no parks and playgrounds for the children. One possible solution here is for the contractor to create a small playground with some benign apparatus to divert the children. He would probably have to donate this to the city to relieve himself of the responsibility for playground accidents, but it might be a solution where no other exists. The work site would also need to be fenced to provide additional security against theft. Guards probably would be required in an area of this sort.

In recent years there has been a spate of incidents involving destruction by social activists who see construction as the destroyer of the environment or a good way to draw attention to a social cause. These protesters have chained themselves to trees and equipment, lain down in front of trucks, put sugar in fuel tanks, and even stolen bulldozers to push over transmission towers. They will be the most difficult to deter, for there is no pattern to their destruction and they are apt to strike from anywhere.

The last category of vandals, union workers attempting to preserve their jurisdiction against nonunion competition, is the most insidious and destructive. J. Leon Altemose, "Construction's Man of the Year" of *Engineering News-Record,* has brought the glare of public disclosure to this disgusting situation. In trying to operate as an open-shop contractor, he has had to fight a virtual war against the labor unions. Besides a physical beating administered by union goons, numerous other incidents created by the union have occurred at his construction site: a mile of 8-ft chain link fence trampled, $300,000 worth of construction equipment burned, and a tractor-trailer rig worth $35,000 destroyed. He is alleged to have spent $667,000 for security and legal expense in 1972 to protect himself from union vandalism.[1]

Contractors are reporting increasing acts of vandalism against their jobs, including equipment on the job. Smashed windshields, ripped-out wires, ruptured radiators, and extensive spray painting (with aerosol-spray cans) are reported with increasing frequency. *Construction Methods and Equipment* reports the annual cost of vandalism as 1 to 6 percent of the value of the equipment.[2] A survey by the Subcontractors Trade Association of New York in January 1971 gave the following average cost of vandalism per contractor:

    1969 $5,333
    1970   6,728   a 25 percent increase

Children and social protagonists will be the easiest to control and deter. Good, tight security measures will usually take care of them. The best solution to union vandalism is for contractor organizations to unify the construction industry to combat this invidious practice of union preservation of the union shop.

PILFERAGE AND THEFT

Pilferage is the stealing of a petty amount of money or trivial objects, while theft is a more general term for the felonious taking of another's property with the intent of depriving him of it, i.e., in contrast to borrowing. The two go hand in hand—after all, it is only a matter of degree and one often leads to the other. Some employees have always considered it a sacred right to pilfer from their employers. It is very difficult to detect pilferage of this sort, because the employee who steals usually takes materials over which he has custody or which he handles regularly. Hand tools,

---

[1] *Eng. News-Rec.,* Feb. 15, 1973.
[2] *Constr. Methods Equip.,* May 1971.

small appliances, stationery, and even small repair parts daily disappear from the warehouse and office. On occasion, much larger items are removed by an employee through some clandestine means, but this sort of larceny is much less frequent than the petty pilferage that goes on. Tote bags, lunch boxes, and packages of any kind may contain the pilferer's loot. Of course, court annals are replete with accounts of embezzlement of funds. These are usually the large amounts detected by an audit or some mistake on the part of the embezzler. Most of the petty cash taken is never detected.

The big question is: How do we stop all this pilferage? There is no simple solution. Many have been tried, but none has ever met with complete success. If the job is fenced, employees' packages may be inspected on departure from the job, but like clever shoplifters, employees may conceal items somewhere on their person. One company which had trouble keeping $\frac{1}{4}$-in electric drills on hand switched to $\frac{1}{4}$-in air-power drills and their losses stopped. This presents a unique solution: if you use things no one can use elsewhere, pilferage of that item stops. Another method of detecting pilferage is to send the suspected employee on vacation or switch his job to another location. An audit or inventory in his absence should reveal any shortages. Every employee should be forced to take at least 2 weeks vacation annually. During this period, accounts can be checked and audits and inventories can be performed without any chance of connivance by the employee.

Theft presents a more serious problem in that large amounts are being stolen with greater frequency. Gasoline tanks are siphoned with great regularity on urban jobs. Sometimes this may be detected by the low miles per gallon recorded on the equipment record. This may be regarded by some as petty, but when large transit-mix trucks are robbed of gasoline regularly, the total monthly loss may be significant. Locking gas-tank caps is one solution. Another may be to switch to diesel or LPG engines. In one large ready-mix concrete company operating in San Francisco, diesel-fuel losses are significantly less than gasoline.

Some firms are being systematically looted by organized gangs of thieves. Losses include not only job materials, but also vehicles, office machines, and large tools and equipment. These represent serious losses, which result in an increased cost of doing work, loss of time, and frustration on the part of management.

Handling security on a large job has become a matter requiring professional attention. There are many techniques

INVENTORY MANAGEMENT

available for security today, some old and some new. Among the things which should be considered to enhance job security are:

Security guards
Watchdogs
Illumination at night
Fences
Infrared or light-beam alarms
Proximity alarms

All of the above are available in the mixed bag of tricks which the professional uses to discourage pilferage and theft.

Another strategem useful to anyone who will take the time is the proper marking or identification of vulnerable items. Tires and batteries, for example, can be branded with the corporate seal in such a way that it can be readily seen and not easily removed. Sometimes things that "walk" can be chained, riveted, or welded to a larger object which cannot be removed. The most important thing is to have some accountability for everything so that anything missing can be detected right away. Awareness and energetic action on the part of supervision and management will go a long way toward the establishment of a climate of security on the construction job. This will tend to discourage all but the more determined thieves.

ACCOUNTABILITY FOR EQUIPMENT

People seldom worry about the accountability for a machine or piece of equipment until something happens to it. Accountability for equipment should be clearly established the day it arrives and not relaxed until final disposal.

CUSTODY OF THE EQUIPMENT

The ultimate responsibility for all equipment, as well as money, personnel, and activity, must rest with the highest executive present on the job. This responsibility must be passed down to those who actually have custody of the equipment. At the lowest end of the spectrum is the equipment operator in the field and the mechanic in the shop. If a machine is assigned to only one operator (highly preferable), then he should be the custodian of the machine. If the machine is operated by others in the same crew or shift, then the lowest supervisor who has sole responsibility for the assignment of operators to the machine should have custody of it. In this regard, it is recommended that whoever has custody of the machine sign a custody receipt for it, this receipt to be held by his supervisor until the machine is reassigned. There is nothing better than a signed receipt to reinforce the local rules on who should be responsible for the care of a machine.

CUSTODY OF PERSONAL TRANSPORTATION

Sedans and pickup trucks are generally assigned as personal transportation, although some may be operated from a transportation pool. In any case, personal transportation must be the responsibility of the person to whom assigned and he should sign a custody receipt for it. Sedans are assigned to managers and pickup trucks are assigned to field supervisors. The latter are probably the biggest cause of transportation problems. The misuse of these vehicles for other than job assignments is very difficult to control. Since they are generally used for home-to-work transportation by the supervisors, they are sometimes found in front of bars on the way to the employee's home. One contractor on the West Coast has a standing rule that any company pickup truck parked near a bar will be picked up by the transportation supervisor and returned to the job compound. It is a management responsibility to maintain a firm hand on the utilization of personal transportation furnished for job use.

ANNUAL INVENTORY OF EQUIPMENT

Equipment should be inventoried annually. At the time of the annual inventory of equipment, each machine should be sighted, inspected for unreported damage, and appraised in market value. This is the time to reaffirm custody of the machine, and the custodian should be present at the time of annual inventory. The annual inventory accomplishes several objectives: it is positive evidence that the equipment is still on hand, it establishes the fact that there is a custodian and identifies him, it updates the inventory record card to show that there has been no unauthorized removal of accessories or unreported accident damage, and it establishes the residual value of the machine.

TRANSFER AND RECEIPT

In order to maintain accountability of equipment, transfer and receipt must be formalized to the extent that all interested parties know about the transactions. When a machine is received, all accompanying documents should be checked to ascertain that the operating manual, parts manual, maintenance records (if used), equipment, and other necessary papers for the proper

INVENTORY MANAGEMENT

utilization of the machine are on hand. If repair parts and consumables such as filters, lubricants, fuels, etc., are not on hand, they must be obtained with dispatch. When a machine is transferred, it should be accompanied by the same information, together with any parts peculiar or parts common not otherwise required, by the sending activity.

THE MANAGER'S RESPONSIBILITY

It is the manager's responsibility to see that policy and procedures for the accountability of equipment are clearly spelled out and disseminated among those who have a need to know. Forms should be developed to formalize established procedures. Many such sample forms are found in the military services and in the private sector of the construction industry. While information may be difficult or even impossible to obtain from the latter, it is easy to obtain from government sources.

One thing is certain about the manager's responsibility in regard to total inventory management and that is no one else is likely to seize the initiative. If he does not take the lead in setting this policy, it probably will not be done.

CONCLUSION

In summary, it is the responsibility of top management to accomplish the following in regard to the management of the equipment inventory:

1. Adopt a classification code for identification of equipment.
2. Establish a registration system and number all eligible machines.
3. Delineate storage policy and establish procedures for both short- and long-term storage.
4. Provide adequate security for the equipment at all times to protect it, especially from theft and vandalism.
5. See that all equipment is assigned to the custody of responsible employees.
6. Assure that an annual inventory of all registered equipment is made.

10
Maintenance Management

Of all the costs related to construction equipment, maintenance costs are the most easily controlled and the most often overlooked. The total cost of maintenance over a lifetime of useful service often exceeds the acquisition cost of the machine. The most probable reasons that these costs do not receive a lot of attention are:

1. They are expected to occur.
2. They accrue slowly, day by day, over a period of several years.
3. They often are not tabulated and brought to the attention of management.
4. They are usually small for routine maintenance.
5. The cost of downtime is often not attached to the maintenance of a machine.

In the context of this book, maintenance includes all labor (both direct and indirect), material, plant, and overhead required to sustain equipment in good, serviceable condition. It includes

MAINTENANCE MANAGEMENT

periodic inspection, lubrication, servicing, repairs, and overhauls.

The secret of a successful maintenance program is to have it systematized in such a way that everyone concerned with both maintenance and operations knows what is expected. This means that there must be a workable method of collecting data such as costs and operating hours (or miles operated), established procedures for preventive maintenance, and a conscientious effort on the part of all employees to implement a system which reduces costs and maximizes profits. The key to the entire program is *scheduled maintenance*. Maintenance can be controlled only if the bulk of it is scheduled. This will reduce downtime and maximize the availability of the equipment to do what it is designed to do—earn money. The implication of a system is regularity and order. When requests for immediate attention become the rule, schedules are forgotten and the organization operates from one crisis to the next. There must be a heavy dependence on operating on a basis of scheduled maintenance or a meaningful system cannot be implemented.

Communication is extremely important in operating such a maintenance system. Management must keep all necessary personnel informed of such things as expected useful life of equipment, replacement policy and procedure, level of decisions which must be made regarding extent of repairs and costs, allocation of equipment, standardization of procedures, need for data collection, and what is expected of each person in terms of performance. On the other side of the coin, workmen must inform supervisors of repeated failures in equipment parts, unfitness of machine to do assigned work as evidenced by breakdown, improper machine operation resulting in parts failure, weakness in machine design, and suggestions for improvement in the maintenance system.

The maintenance organization has an obligation to establish the schedules for equipment and to keep the operating departments advised of these schedules. This scheduling must be accomplished with top management delineating the maintenance philosophy. Maintenance should be scheduled to minimize the effect of any downtime on regularly scheduled operations. Operating divisions should keep the maintenance supervisor informed of any inability to comply with the schedule established, so that the maintenance can be rescheduled at an early, more convenient time.

SCOPE OF A MAINTANANCE SYSTEM

A maintenance system should include provision for all degrees of maintenance. These would include at least the following:

1. Preventive maintenance
2. Repairs
    a. Minor or field
    b. Major or shop
3. Overhauls

Not all of these maintenance functions are necessarily performed by the owner or operator of the equipment. Major repairs and overhauls are often done in the dealer's shop or at some commercial establishment. For highway equipment operating far from home base, even preventive maintenance may be handled at commercial garages.

Procedures must be established to carry out the necessary maintenance on a piece of equipment from acquisition to disposition. Basic to this process is the design or adoption of forms to accomplish these procedures in an orderly fashion with only the minimum paperwork required to achieve the objective. Where people are required to record data, they should be recorded only once if possible. Let the computer or data-processing equipment do the rerecording, shuffling, and manipulation necessary to get out the reports. Remember that as a minimum, you must record the annual cost of routine maintenance, one-time cost of major repairs and overhauls, and the operating hours related to these costs. When the machine is finally sold, you should have the total cost of all maintenance by years and the associated operating hours. This is the minimum information required. Most owners will want to keep more records than these. The surprising fact is that many contractors do not keep even the costs mentioned here.

Most contractors agree to the separation of costs into labor, parts, tires, lubrication and preventive maintenance, accident repairs, and special modifications. Further breakdown should be based on a careful weighing of costs versus benefits to be derived from them. The quantification of benefits is difficult and subjective and hence will reflect the prejudices and preferences of the person making the analysis. One thing is certain: the more raw data collected, the more money must be spent to screen them for reliability. Persons recording the data must understand why they are being collected and the importance of accuracy. Two-way communication between the person designing the data form and the mechanic will go a long way toward solving this problem. If the data are useful to the mechanic too, their accuracy will be increased. It is highly preferable that data be recorded as in double-entry bookkeeping, so that their accuracy can be determined by periodically balancing the books. For example, if the number of gallons of gasoline consumed by all vehicles is totaled at the end of the month, it should equal the total gallons dispensed at the fuel depot (or tank truck). The latter amount can be reconciled with the amount on hand at the beginning and end of the month and the amount purchased.

One practice that must be guarded against in reporting maintenance costs is the fudging or distortion of raw data at the source. This may be done by the mechanic to cover up personal mistakes, to justify past decisions or reinforce arguments about the high cost of certain machines, because of carelessness or laziness, or just to harass the supervisor. Whatever the cause of inaccurate data, they are the primary source of unreliable cost records and the principal reason why most contractors shy away from a complete and total accounting of cost.

The method of record keeping should be simple and clear and have enough checks and balances to be reliable. Basic to this is the design or adoption of a good set of reporting forms. These forms should provide for the signatures of two people, the employee recording the data and his supervisor who checks them. For even a minimal system, at least the following forms will be required:

1. Equipment repair history
2. Shop repair order
3. Preventive-maintenance record
4. Service-station report (fuel and other services)
5. Daily operator inspection report
6. Operator report of equipment deficiency

Standard forms for these reports are printed by many printers and suppliers of business forms. If the contractor wants to design his own, guidance may be obtained from several publications.[1]

[1] For example, Management of Transportation Equipment, *Nav. Facil. Eng. Command Man.* P-300, September 1971; Bernard T. Lewis and James P. Marron, "Management of Vehicular Operations and Maintenance," Rider, New York, 1965.

Experience of those in the industry indicates that a manual system is satisfactory for an inventory of fewer than 500 machines. Above that number, a computerized system is more economical. The principal determinant here is labor. It takes a minimum of three people to operate a computerized data-processing system, and that is the number of persons required to handle 500 machines in a manual system. The ease of retrieval of information and the facility for manipulation of raw data by the computer have caused management to rely less on intuition and more on informed analysis as a method of making equipment decisions. It is probable that in the future, canned computer programs for equipment costing and analysis will be made available to small contractors, so that fleets of fewer than 500 machines can be managed economically through computer methods rather than manually. The final test on this will be the cost of labor required to collate the raw data for analysis.

PREVENTIVE MAINTENANCE

The first line of defense in a good system of maintenance is a well-planned program of preventive maintenance (PM). Preventive maintenance includes but may not be limited to:

1. Inspection (includes checking and testing)
2. Lubrication and greasing
3. Replenishment of consumables (filters, etc.)
4. Servicing (air, battery water, coolant, etc.)

All of the above are accomplished at periodic intervals in accordance with the manufacturer's recommendations or the user's experience. The former may be used with a new machine and then adjusted by the user in light of his application of the machine and working conditions. The interval between maintenance operations may be determined in any one of three ways: calendar time, hours of actual operation, or miles traveled. Usually the interval will be determined by calendar time in combination with operating hours (heavy equipment) or miles traveled (highway equipment). In carrying out an adequate preventive maintenance program it is absolutely essential that odometers and hour meters operate. Often they are considered to be one of the less necessary functioning parts of a machine, but this is far from the truth.

Periodic maintenance can be operated in one of three ways. It can be scheduled in periods recommended by the manufacturer for average service conditions, in which case it is called

MAINTENANCE MANAGEMENT

standard-interval maintenance. For example, for a highway dump truck the manufacturer might recommend greasing the front-wheel bearings at intervals of 10,000 mi. If this were set up on the preventive-maintenance (PM) program, front-wheel bearings would be repacked every 10,000 miles. This would be standard-interval maintenance.

The next refinement above standard-interval maintenance is called *extended-interval maintenance.* Here the interval is adjusted by the user to match his application of the equipment: severity of operations, ambient temperature, operator proficiency, dust, and other conditions of work. Suppose in the case of the dump truck cited above that the contractor never had any wheel-bearing failures. He might then extend the interval by 2,000-mi increments until he started to get a few bearing failures at 18,000-mi intervals. In this case, he might adopt a 16,000-mi interval for packing wheel bearings. This 16,000-mi period would be a use of extended-interval maintenance.

The ultimate refinement in periodic maintenance is *controlled maintenance,* in which the interval is adjusted each time the operation is performed in accordance with the needs of the machine. Using the example above, consider the front-wheel bearing requirement for greasing under a controlled-maintenance program. The bearing would be monitored either by inspection or by some detection device and would not be greased until required. This might be at 16,000 mi the first time, 18,000 mi the second time, 15,000 mi the third time, and so on.

There are several types of PM checks, classed according to the intervals between them. They are commonly classed as follows:

	Interval	
Type	*Vehicles*	*Construction equipment*
Operator	Daily	Daily
A	2,000 mi or 1 month	100 hours or 1 month
B	10,000 mi or 6 months	500 hours or 6 months
C	20,000 mi or 12 months	1,000 hours or 12 months

These intervals are not mandatory but are adjusted according to the job conditions. A safety inspection is generally included in the type B inspection and should be performed at least every 6 months.

The most frequent PM is performed daily by the operator.

He should check his machine before start-up in the same way an airplane is preflighted before takeoff. The importance of this check should be emphasized: it is the most important one of all. Before starting the engine, the operator must check at least the following:

1. Air pressure in tires
2. Fuel level
3. Crankcase oil level
4. Coolant level
5. Battery water
6. Leaks
7. Any obviously broken or defective parts noted by walking around the machine
8. Safety items (brakes, lights, horn, etc.)

Other items will need to be checked by the operator also, depending on the type of equipment and job conditions. Any deficiencies should be reported on the operator's daily check form and immediately to his supervisor if safe operation of the equipment is involved.

    The type A PM is performed either in the field or in the shop. It involves the most frequently performed inspection and maintenance. This includes oil changes, filter cleaning or replacement, adjustment of drive belts, tire check, and other routine operations. Consumables such as antifreeze, hydraulic fluid, transmission oil, and the like are replenished.

    The type B and C PMs are usually performed in the shop except for large pieces of heavy equipment. The inspection is much more extensive than the type A inspection, and such things as wear on clutch facings, brake bands, bearings, and other parts should be checked.

    In order to implement a good program of preventive maintenance properly, each piece of equipment should have a list of items to be checked on each type of PM. Also included for each machine should be a diagram showing the location of all grease points which must be lubricated at each interval. Some companies put these checklists and diagrams in a book form with plastic binder. The type A, B, and C lists are interspersed in the proper order in the book. In this manner a permanent file is kept of all PMs, especially important on all large and expensive machines.

MAINTENANCE MANAGEMENT

A word needs to be said about maintenance personnel. PM is the backbone of your entire maintenance program. The grease rack and tire shop should not be staffed with cast-off, incompetent mechanics who are not considered good enough to work in the repair shop. One of the reasons for this practice is that grease racks and tire shops often are dirty, untidy, and uninspiring. Men are sent there as a punishment for their shortcomings. This need not be so. Have you ever seen a filthy sewage treatment plant? Never! Maintenance shops can be staffed with lubrication experts, cleaned up and painted bright colors, and piped with sound. This will reflect very soon in the performance of the equipment. Good maintenance must start with good, conscientious employees.

There are several ways by which deficiencies in equipment are found. They may be reported by an operator as the result of his inspection or observation during the working day; they may be detected at a regular PM inspection; they may cause the machine to break down in the field; and they may be observed after an accident. Most deficiencies detected by inspection can be corrected by scheduling the repair so that downtime is minimized. Breakdowns and accidents usually result in unscheduled maintenance and hence more downtime and higher cost. Naturally, scheduled maintenance is preferred.

SCHEDULED MAINTENANCE

In a well-run equipment-maintenance operation, it is usually estimated that about 75 percent of the maintenance will be scheduled and 25 percent unscheduled. The bulk of the scheduled maintenance will result from the preventive-maintenance program.

How much downtime is associated with maintenance depends on several factors. There will be less downtime associated with scheduled maintenance because repair parts will be on hand before the machine is scheduled for the work. If a job is on a single daytime shift operation and the shops work the same shift, a reasonable estimate of downtime is considered to be 7 percent of the scheduled working hours. In other words, on a single-shift basis of 2,000 scheduled hours per year, a machine will average about 140 hours of downtime. This downtime can be broken down further into 2 percent awaiting repair, 2 percent under repair, and 3 percent awaiting parts. If preventive maintenance and repairs on a single-shift operation are scheduled outside of regular-shift working hours, say on a night shift, downtime can be reduced to less than 5 percent.

Cleaning and painting of equipment are important adjuncts to a good maintenance program. Equipment should always be cleaned before a PM check. Dirt and mud not only cover up deficiences and broken parts but hide grease points and parts to be lubricated. A grease gun placed on a dirty fitting will force grit and other foreign matter in, to contaminate a bearing or wear surface. Cleaning should preferably be accomplished with high-pressure water. If a steam jet is used at all, it must be handled with care to avoid melting the grease out of wheel bearings and other lubricated surfaces. Engines should be degreased periodically to enhance servicing and reduce the fire hazard incidental to accumulated petroleum products.

Painting of equipment is necessary to engender pride as well as to preserve the machine. A machine which looks well will usually be treated accordingly. Not to be forgotten in the appearance of equipment is the public-relations aspect. People who see an attractive, well-cared-for machine will naturally assume the contractor handles all his business that way. And last, but not least, a clean machine with a highly visible paint job is a safer machine. Machines should not be painted earth colors unless it is necessary to camouflage them. International orange is the highly visible color used for obstruction painting around airfields and is a good one for equipment, too. Most light colors will do but some, like white, will soil easily. Yellow or orange appear to be the best choices.

Scheduled repairs on equipment can be accomplished in a variety of ways. Most minor repairs can be done in the field but may be done in the shop if it is nearby. The contractor usually has his own field-repair crew who work out of the repair shop on the project. The repair shop on the project is generally geared to do all preventive maintenance, minor repairs, and possibly some major repairs. Beyond that, the contractor's centralized repair shops will accomplish any major repairs above the capability of the field shops and any overhauls or component rebuilding that is desirable. With the advent of fast transportation (helicopters, etc.) and rapid communications, many components are purchased from the dealer on an exchange basis. The dealer then remanufactures the bad component and replaces it in his exchange stock.

Some repair or overhaul work is accomplished on a contract basis by commercial shops or specialty contractors. Tire maintenance, for example, is often subcontracted to a dealer or representative of the manufacturer. Major engine overhauls are frequently performed in the local dealer's shop. There are some advantages to this method of getting major work done. The specialist usually has experts with more know-how than the average contractor. As a result, the work is better and more economically done. Sometimes a small contractor does not have enough clout to get sufficient priority with a large dealer, especially if a larger customer is waiting in line. For this reason, many contractors like to do their own major repair work, as they have better control of the scheduling.

UNSCHEDULED MAINTENANCE

No one wants unscheduled maintenance, but it is not all preventable. Most of it should result from operator-reported deficiencies in between type A PM checks. Unfortunately, however, some will result from breakdowns and accidents. Most of the two latter causes can be reduced by alert management, and with proper planning their effects can be alleviated. In the section on scheduled maintenance above, an average of 7 percent downtime was mentioned. In some operations, downtimes of 15 to 20 percent are encountered. Downtimes of this magnitude should be accepted only after the most careful weighing of all pertinent factors.

It has been a practice among some contractors to operate on a breakdown-maintenance policy rather than one which includes preventive maintenance. Ordinary lubrication will be accomplished at unspecified intervals, but the machine never receives any shop attention unless it breaks down on the job. In this case an emergency repair is necessary, especially if it is a key machine. This disrupts shop work, robs management and supervision of valuable time, and increases cost in the long run. If incipient breakdowns cannot be detected by timely inspection, management should determine what is wrong with their maintenance policy.

One cause of breakdowns and accidents is the misapplication of a machine in the field. Supervisors are almost always responsible for this turn of events. Weight-lifting equipment is especially susceptible: machines overturn, booms bend, and parts break, all because someone did not check on the capacity of the machine or accurately calculate the weight of the load.

Even worse than misapplication is misassignment of equipment. This generally goes above supervisor level and rests on management. Using rubber-tired equipment where crawler machines should have been used can result in catastrophic tire cost. There are many examples in everyday operations where the wrong equipment or even the wrong design of the correct machine has ended up by costing the contractor a mint. Failure to include these costs in the bid is due to the contractor's failure to recognize the conditions and account for them. Careful planning in advance will help to reduce the risk of high maintenance costs resulting from use of the wrong equipment.

Sometimes a reliability ratio is used to measure the effectiveness of a maintenance program. The interested reader should refer to Lewis and Marron[1] for a more complete discussion of reliability and its relationship to maintenance costs. The reliability ratio is defined as

$$R = \frac{S}{U}$$

where R = reliability ratio
S = scheduled maintenance (PM inspections, scheduled repairs, etc.)
U = unscheduled maintenance (breakdowns, accidents, etc.)

Each visit counts as 1, and it is assumed equal time is expended in each visit, scheduled or unscheduled. The minimum acceptable R is equal to 3. This can be imputed from the fact that an average of 75 percent scheduled maintenance divided by 25 percent unscheduled maintenance equals 3. The fewer the visits for unscheduled maintenance, the higher the reliability ratio will be. This is a rather gross measure of the effectiveness of the maintenance program. An R of 2 would be unacceptable and an indication that steps for improvement are mandatory. The reliability ratio will be influenced by at least the following factors:

1. Competence of mechanics
2. Competance of operators
3. Careful attention to schedules for preventive maintenance
4. Severity of job conditions
5. Quality of equipment

It is recognized that the reliability ratio is an imperfect way to gauge reliability, yet it does provide some guidance where there is none.

[1] Op. cit., pp. 125ff.

MAINTENANCE MANAGEMENT

MANUAL SYSTEM FOR SCHEDULING PREVENTIVE MAINTENANCE

As mentioned in a previous section, up to 500 pieces of equipment can be controlled by manual methods. Such a system is fully described in a Navy manual,[1] but a brief description of the method will be given here.

Obtain a file drawer or rectangular box approximately 12 in wide and 24 in deep. If a box is used, it may be mounted on locking wheels so it can be moved to a convenient location for sorting. Figure 10-1 shows how a box might be set up. Next, get 126 file folders or spacers and number them from 1 to 126, arranged consecutively in the box. These separate the PM record cards according to service intervals as explained next. The reason only 126 groups are needed is that there are 252 working days in a year and 126 days, or 6 months, is the maximum interval for type A service.

The next task is to fill in a 5- by 8-in PM record card similar to the sample shown in Fig. 10-2 for each piece of equipment. Calculate the PM group number for each machine. The PM group (PMG) number is the number of days intervals between consecutive type A services. It can be calculated by using the equation

$$\text{PMG} = \frac{252 \times \text{SI}}{\text{AW}}$$

where PMG = PM group
SI = service interval in operating hours (equipment) or miles (vehicles)
AW = annual work in operating hours or miles

Suppose, for example, a crane has type A PMs at 100-hour intervals and accrues 1,500 operating hours per year. Then

$$\text{PMG} = \frac{252 \times \text{SI}}{\text{AW}} = \frac{252 \times 100}{1,500} = 16.8, \text{ say } 17$$

This crane would be listed as PMG = 17, or type A service would be required every 17 working days.

After all machines have been listed on cards and the PMGs have been calculated, they are sorted by PM groups. Now count the number of machines in each group and tabulate them. The number of days between services for machines in each group will be calculated next from

$$D = \frac{\text{PMG}}{N}$$

---

[1] *Management of Transportation Equipment, Nav. Fac. Man.* P-300, Naval Facilities Engineering Command, September 1971.

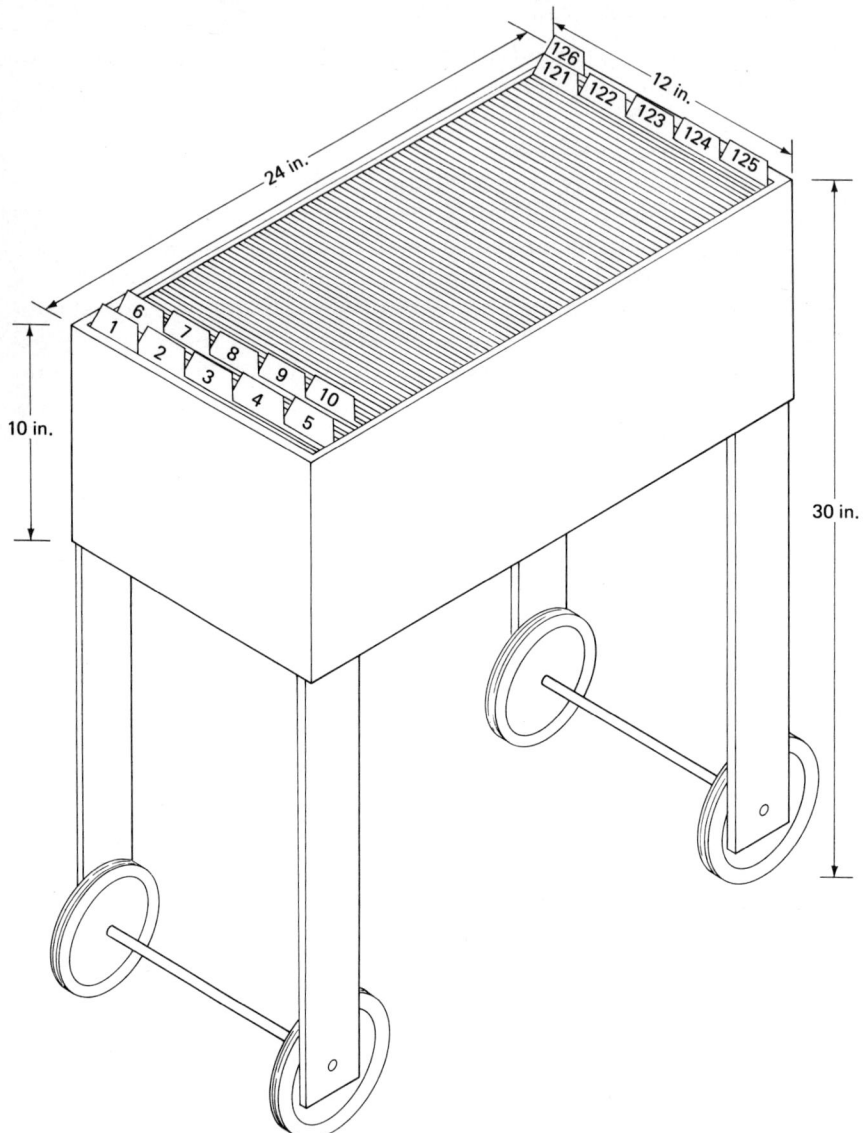

Fig. 10-1 Box for sorting PM cards.

where D = days between services for machines in PMG
PMG = PM group
N = number of machines in group

MAINTENANCE MANAGEMENT

1. Equipment number				2. Equipment type			3. Project number		4. PM Group	
5. Manufacturer				6. Model			7. Year		8. Estimated annual op. hrs. (miles)	
9. Type PM	10. Date	11. Cumulative op. hrs. (miles)	12. Op. hrs. (miles) since last PM	13. Op. hrs. (miles) this 6 mo. period	Type PM	Date	Cumulative op. hrs. (miles)	Op. hrs. (miles) since last PM	Op. hrs. (miles) this 6 mo. period
			Entries from last record						
*			Last semi-annual entry						
*			Op. hrs. (miles) at transfer						

*Enter only on transfer

Fig. 10-2 Sample PM record card.

For a contractor operating 175 pieces of equipment, the computation would be as follows:

PMG	N	D	Quota
126	62	126/62 =	2.0 or 1 every 2 days
105	12	105/12 =	8.8 or 5 in 44 days
92	8	92/8 =	11.5 or alternating 11 and 12 days
63	35	63/35 =	1.8 or 5 in 9 days
42	4	42/4 =	10.5 or alternating 10 and 11 days
28	36	28/36 =	0.8 or 5 in 4 days
17	18	17/18 =	0.9 or 10 in 9 days

Since the interval does not always turn out to be an even number of days, the quota may be obtained by taking the reciprocal of D. For example, PM group number 63, which has 35 vehicles, has an interval $D = 1.8$ days. Since the reciprocal of 1.8 is 5/9, you will need to schedule five machines every 9 days. This could be accomplished by scheduling the first 4 at 2-day intervals and the fifth 1 day later.

Now the PM record cards can be distributed in the file folders according to the intervals determined above. Start with PMG = 126 and place one card each in days 1, 3, 5, 7, ... at 2-day intervals until the last is filed in folder 125. Next distribute the cards in PMG = 105 placing them in folders about 9 days apart so that you schedule five every 44 days. Level out the work load and try to schedule approximately the same number of machines for PM each day until all 175 pieces of equipment are scheduled. Adjusting the schedule a day or two to accommodate the work will make no difference. A calendar should now be prepared to cross-reference the working days shown in the PM file.

Equipment can now be scheduled directly from the file folders. On day 1 all vehicles or equipment whose record cards are in that file should be serviced. Upon completion of servicing, the fact should be recorded on the PM record card of the machine. The file folder is now removed from the front of the file and placed in the back. All cards in it are redistributed according to the PM group or interval to next service. For example, suppose file 1 has one PMG 126 card, one PMG 92 card, two PMG 28 cards, and one PMG 17 card in it. At the conclusion of servicing and notation of records, file 1 is placed to the rear behind file 126 with one PMG 126 card in it. One PMG 92 card is moved to file 93, two PMG 28 cards are moved to file 29, and one PMG 17 card is moved to file 18. In other words, the PMG number is added to the old file number to obtain the address of the new file.

This is the basic plan of a manual-scheduling system for preventive maintenance. It will be less complex for a small number of machines and once mastered can easily be handled by one person as a part-time assignment. This same scheduling process has been computerized for large fleets, and several oil companies offer their scheduling plans to customers in order to stimulate sales.

INSPECTION OF EQUIPMENT

Inspection is an integral part of the preventive-maintenance program, but its importance is often overlooked. The purpose of inspection is to detect failure before breakdowns or accidents occur. It is intended to minimize unscheduled repairs and enhance the safe operation of the machine. Mature judgment is one of the most important characteristics of a good inspector. He must have sufficient training and broad experience in order to separate the important things he observes from those of less importance. It is not enough that he be a good mechanic; he must be able to project present conditions into future operations and forecast any dangerous outcome. He must be able to answer such questions as: will a hairline crack in the frame of an important piece of equipment be reasonably safe if the machine continues operation, or must the machine be deadlined now for repair? The aim of good maintenance is to keep the equipment in serviceable rather than perfect condition, and the inspector is a key man in this plan.

To aid the inspector in his work, modern technology has provided an ever-increasing array of sophisticated diagnostic equipment and techniques. A few of the most important will be mentioned here; many others are available. Lubricating-oil analysis, chassis dynamometers, ignition analyzers, wear detectors, and automatic lubrication are some of the things that are in use today. Tomorrow's tools and techniques will soon be with us.

Lubricating-oil analysis This technique has been commercially available since the middle fifties. It enables the mechanic to diagnose an incipient failure without tearing down a component. It is equally applicable to any type of engine, transmission, final drive, differential, compressor, hydraulic, or closed lube system.

Several oil companies perform the tests free of charge for their customers, while commercial laboratories charge for the service. In 1972 the Caterpillar Tractor Company made available to the industry a program of Scheduled Oil Sampling (SOS). This service can now be obtained through many of its dealers and will be used to exemplify the techniques.

A prepackaged sampling kit consisting of a sampling gun, plastic tubing for drawing samples, sample bottles, and mailing tubes is available from the dealer on loan. Replacement items are available at a nominal charge. The customer takes samples of lubricating oil from the engine crankcase, transmission, and final drive of the machine to be checked. These samples are sent in to the dealer for testing. Samples which arrive before 9:00 A.M. will be tested that day and results mailed to the customer by 4:30 P.M., providing same-day service. The service is free to customers, and may include testing of other than Caterpillar equipment.

Samples from the three points mentioned are taken at periodic intervals, say 100 hours as a start. At least three successive samples are required to establish a norm for the equipment; nevertheless, early samples are indicative of existing conditions. The purpose of the analysis is to determine the inventory of wear particles and foreign matter in the lubricants. The sample is tested for three things, namely, water and ethylene glycol (antifreeze) contamination, fuel dilution, and wear particles and foreign matter. The first two tests are straightforward chemical tests. The tests for wear particles and foreign matter are performed by an atomic absorption spectrophotometer.

The last tests are performed in batches, which considerably accelerates the process. A small amount of oil is burned in a hot gas flame and the amount of the trace element is measured in parts per million (ppm) by the apparatus. The elements generally measured and their significance in the sample are as follows:

Element	Compartment	Indicated problem
Silicon	All	Entry of dirt
Chrome	Engine	Wear on rings, crankshaft, valves
Copper	Transmission	Clutch disk or bushing
Iron	All	Gears, splines, bearings
Aluminum	Engine	Pistons, bushings, bearings

Lead, silver, tin, and other trace elements are measured if there is any significance in their presence. As a matter of fact, it has been suggested that different trace elements be included in the metallurgy of the different parts to assist in locating unusual wear.

In the SOS program, normal wear limits are established for the engine, transmission, and final drive as indicated by the presence of the measured elements. As a rule, the more operating hours on a lubricant, the more wear particles it will contain. Time to change to oil will be indicated, as well as unusual wear. General wear limits for each compartment of each model have been established as shown in the following example for a D8 crawler tractor engine:

Element	Amount, ppm			
	Acceptable	Reportable	Unacceptable	Urgent
Fe	< 45	45–70	71–95	> 95
Cu	< 10	10–15	16–30	
Al	< 9	9–15	16–18	> 18
Cr	< 4	4–12	13–20	> 20
Si	< 15	16–23	24–30	> 30

Items which are unacceptable should be investigated and corrected. Any item within the urgent category is the subject of immediate telephone report to the equipment owner, as it indicates impending failure. More information on this program can be obtained from the local Caterpillar dealer.

A good oil-analysis program will accomplish the following:

1. It warns of inadequate maintenance. The presence of silicon, for example, may indicate improper air-cleaner maintenance.
2. Life of components is increased by immediate detection of unusual wear. This enables the owner to schedule his downtime for repair.
3. Repairs and adjustments can be made before a major failure occurs, thus reducing total maintenance costs.
4. Availability of machines is increased, thus increasing productivity and efficiency.
5. The contractor has a better knowledge of the condition of his equipment, enabling him to plan better.
6. Service intervals generally may be increased, reducing cost of maintenance.

Although lubricating-oil analysis has been around for a good many years, the bulk of the users appear to be large organizations such as the Navy, Air Force, major airlines, larger trucking fleets, and large corporations owning equipment. L. S. Dryden,[1] of the General Electric Company, reports that on the basis of laboratory analyses of used oil, oil drain periods have been increased in corporate equipment from 3,000 to 9,000 mi. There is no doubt that a comprehensive program of oil testing will result in many benefits to the equipment owner and user.

Dynamometer This is a device for measuring engine horsepower at various engine speeds. Both electric and hydraulic dynamometers have been available for many years for laboratory testing of engines. In the past decade or two, the rapid development of the hydraulic-absorption dynamometer has lowered the cost of this device to make it a practical tool for diagnosis and adjustment in a well-equipped shop. Two types are generally available,

---

[1]L. S. Dryden, Computer Keeps Careful Check on Vehicle Maintenance Costs, *Plant Eng.*, August, 1967.

chassis and engine. The chassis dynamometer is used for testing operational equipment mounted on a chassis, while the engine dynamometer is used on the separate engine for run-in preinstallation testing.

The chassis dynamometer is most useful for maintenance and will be described here. It consists of a set of drive rolls which support the drive wheels of the truck or vehicle (they cannot be used for crawler equipment). If the vehicle has tandem axles, a set of bogies may be added to accommodate them. The horsepower taken from the drive wheels is varied by adjusting the amount of water in the dynamometer. In order to evaluate results, an instrument panel with readings of horsepower and vehicle speed in miles per hour is visible to the mechanic. Also available as optional panel instruments are an engine tachometer, cam dwell meter, vacuum gauge, and exhaust-gas analyzer.

In testing, the horsepower at various speeds can be checked. Adjustments of the fuel system, ignition system, automatic

Fig. 10-3 Modern chassis dynamometer with brake analyzer. *(Clayton Manufacturing Co.)*

MAINTENANCE MANAGEMENT 211

transmission, or other running gear are simply and directly accomplished by peaking the horsepower at the desired speed. Almost any road condition such as steep grades, heavy or light pulls, and low or high speed can be simulated and repeated exactly when desired.

Dynamometers are not cheap. The installed cost of a good chassis dynamometer may run in the neighborhood of $20,000. The final cost, of course, will depend on the type, size, and elaborateness of the setup. Supervisors who have them, however, agree that they are useful and versatile money savers if the fleet is big enough to justify them economically.

Ignition analyzer These devices are designed around a cathode-ray oscilloscope. They are used on the ignition system of an Otto cycle (gasoline) engine to diagnose problems in the spark plugs, distributor, coil, and other components of the system.

Figure 10-4 shows a 23-in oscilloscope combined with other test instruments to give a wide range of information, including exhaust-gas analysis of hydrocarbons and carbon monoxide. The

Fig. 10-4 Modern ignition analyzer combined with other test equipment.

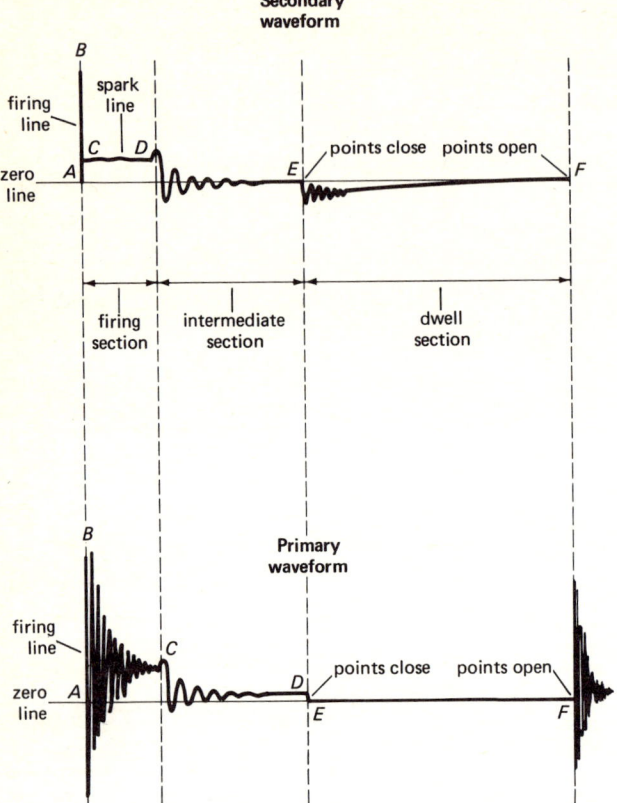

Fig. 10-5 Wave patterns on ignition analyzer.

oscilloscope may be set to display either the primary or secondary waveform of the ignition system, thus enabling the mechanic to diagnose many problems associated with it. The pattern may be displayed in a variety of ways depending on the choice of the operator. All cylinders can be viewed individually or their patterns can be combined in several ways, namely, superimposed, stacked, or in parade. Figure 10-5 shows this selection of patterns for several of the possible tests.

Figure 10-6 illustrates the pattern of the secondary waveform, showing the sequential actions that take place in the firing of a cylinder. Polarity, dwell, cam-lobe accuracy, coil, condenser, points, and spark plugs can be analyzed by proper use of the oscilloscope. The ignition analyzer and associated instruments can

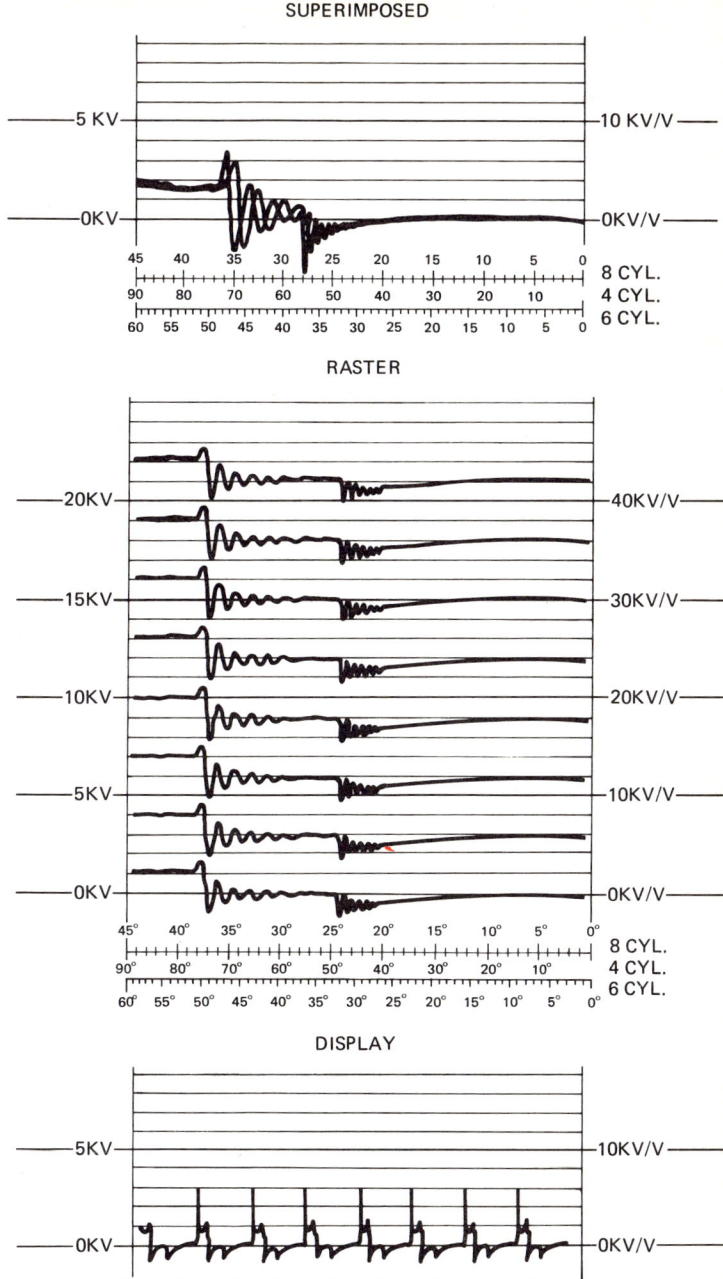

Fig. 10-6 Secondary waveforms for 8-cylinder engine. *(Sun Electric Corp.)*

be combined effectively with the chassis dynamometer and brake analyzer to give a very complete set of diagnostic equipment.

Automatic lubrication systems These have been around since the early 1930s, when automobiles were first equipped with a centralized lubrication system actuated by a push button on the dashboard. Today both Alemite and Lincoln Engineering manufacture systems which can be installed on almost any machine at a very modest cost (about $400 depending on the number and types of machines, subject to considerable variation).

These lubrication systems can be actuated by various mechanisms: clock, total revolutions (of say a wheel or crankshaft), and repetitions (of say a clutch or brake pedal). It would seem wise to consider installation of automatic lubers on all expensive equipment where feasible.

Results of automatic lubrication are dramatic; some equipment users report triple or quadruple bearing life on vulnerable parts, others report no failures where failures were common before installation. The best time to opt for automatic lubrication is, of course, while the equipment is being manufactured.

Wear detectors These are somewhat innovative at present, and their use is being gradually developed with improved technology. Two types are principally under development, electrical and mechanical. Automobiles are now on the market, for example, which have an "idiot light" which flashes on red when clutch facings are worn down. Under development are telltales for alerting the operator on such things as worn brake bands and other fast-wearing parts. Devices are also being used to alert the operator when the air filter needs cleaning or other filter elements need changing. Still needed are reliable detectors for tire air pressure, battery condition, and coolant level.

OTHER PROBLEMS IN MAINTENANCE

Several other problems in maintenance are important enough to deserve some discussion. They are common to many types of contractors and over all geographic areas. These are cannibalization of parts, abuse of vehicles and equipment, an attitude toward machine design which does not include ease of maintenance, and competence of mechanics not matched with the complexity of the machine.

Cannibalization By cannibalization of parts is meant the removal of parts from a deadlined machine or one of low operational priority to restore another machine to operating condition. This practice is more common on remote jobs or those where parts support has broken down. It is a costly procedure and more than doubles the repair labor cost, for the part must be removed from one machine to place it in another. A second repair job is thus created, for sooner or later the part must be replaced in the first machine.

More often than not, replacement of the originally removed part never takes place. The deadlined machine becomes a prime candidate for parts supply and finally its skeleton is abandoned to rust away in ignominy. Cannibalization is not to be condoned except in the most desperate circumstances. When machines are cannibalized for parts, they must not be left open to the weather. Engine outlets must be sealed and the machine otherwise prepared for long-term storage.

Abuse Vehicle and equipment abuse and misuse are the cause of increased repair costs and unavailability of equipment. *Air Force Manual* 66-12[1] lists the following examples of vehicle abuse:

1. Tampering with governors to allow overspeeding.
2. Running engines at excessive speeds unnecessarily and before normal operating temperatures are reached.
3. Operating vehicles with insufficient oil or coolants.
4. Failure of drivers to report malfunctions, defects, and damages affecting mechanical condition and safe operation.
5. Riding or slipping clutches except when necessary to maintain control of a vehicle during backing operations.
6. Operation of vehicles in improperly selected gears; lugging in high gear; use of reverse when travelling forward.
7. Excessive use of engines for braking or overspeeding of engines when used to assist in braking.
8. Operation of vehicles with broken tire chain links or improperly inflated tires.

---

[1] Vehicle Management and Maintenance, *Air Force Man.* 66-12, June 1961.

9. Overloading, improperly securing or improperly distributing loads over cargo area of vehicles.

To the above can be added several more:

10. Unnecessarily skidding or sliding tires, either by braking or turning.
11. Driving too fast for road conditions ("cowboying").
12. Operating a large pneumatic-tired vehicle (especially a truck crane) on a soft, unstable roadway so that it overturns.

Others may be added as the manager draws on his own experience and conditions of doing work in his kind of business. All equipment operators and vehicle drivers should be cautioned about vehicle abuse, and prompt disciplinary action must be taken to curtail it.

Ease of maintenance For too long the attitude toward equipment design has encouraged crowding too much in too little space, locating in inaccessible places parts which need to be seen or often replaced, and fastening parts with semipermanent attachments when easily removable attachments are needed. Ease of maintenance is defined as a measure of the effort, time, and skill required properly to service, adjust, and repair machinery.

Some years ago the Army's Engineer Research and Development Laboratories at Fort Belvoir undertook a program designed to improve the ease of maintenance. One innovative development of this program is the hydraulic track adjuster now used on crawler tractors.

Some of the design principles which should be followed to improve ease of maintenance are:[1]

1. High mortality parts, components, and major assemblies should be accessible for repair and replacement with minimum removal of other assemblies or components.
2. Performance of preventive maintenance should be so easy and obvious to the operator that he performs all day-to-day maintenance functions almost automatically.

---

[1] Ease of Maintenance, *Eng. Res. Dev.* Pam. Corps of Engineers, U.S. Army, Fort Belvoir, Va., circa 1960.

3. Designs should provide easy servicing, adjustment, and repair. Guesswork on the part of the operator, service man, or mechanic should be minimized.
4. All connectors should be standardized and the number of sizes held to a minimum. Standard tools should be adequate for servicing, adjustment, and repair of equipment.
5. Lubrication points should be located in accessible places and obviously identified to simplify servicing.
6. The number of lubricants and fluids required should be minimized and standardized. As a design objective they should be limited to one each: engine lubricating oil, extreme-pressure gear lube, hydraulic fluid, and grease.
7. Drain and check plugs should be of the same type and standard size.
8. Covers, guards, and housings should be rapidly and easily removable. Reduction of the number of bolts in engine hoods and battery covers and the use of hinged radiator guards and housings should be stressed.
9. Mechanical, hydraulic, and electric connections should be quick-release fastening devices except on safety systems such as brake lines. On fluid lines, these devices should have self-sealing features to prevent loss of fluid.
10. The power package including the cooling system, engine, and clutch should be replaceable as one unit and should incorporate standardized mounting.
11. Lifting eyes or threaded stud wells should be provided on all heavy assemblies.
12. Dowels and guides should be provided to aid in aligning heavy, bulky assemblies.
13. Special attention should be given to the selection of bolts, studs, cap screws, and fastening devices which will optimize ease of maintenance.

In order to implement the ease of maintenance program, the Corps studied certain repairs which might be made to a motor grader of commercial manufacture. Table 10-1 shows the fantastic reduction in man-hours achieved by concentrated attention on fasteners, connectors, and other features to enhance the repairs.[1]

---

[1]Ibid. p. 10.

Table 10-1 Time study improved maintenance

Operation	Man-hours			Reduction %
	Before	After	Saved	
Remove engine	3.8	0.5	3.3	87
Install engine	4.1	0.8	3.3	80
Remove transmission	7.6	1.6	6.0	79
Install transmission	11.7	2.4	9.3	79
Remove front-wheel hub and shaft	2.0	0.9	1.1	55
Install front-wheel hub and shaft	2.1	0.6	1.5	71
Remove brake	3.3	0.5	2.8	85
Install brake	2.9	0.4	2.5	86

The results of the Corps of Engineers study are impressive. Contractors should insist that manufacturers pay heed to the basic rules for ease of maintenance outlined herein. Whenever an opportunity is presented for the contractor's maintenance and repair shop to augment these design principles, it should be established policy to do so.

Matching mechanics and machines It is not easy to find mechanics competent in all phases of repair work. Engines, transmissions, and accessories are so complex on today's machines that only a highly trained specialist is qualified to work on certain components. It is well to leave these repairs to the distributor or some factory specialist who can accomplish a good repair.

A repair improperly done is sure to return to haunt the shop. Many such returns for redundant repair escape the attention of the foreman mechanic. This is where a good reporting and accounting system which reports by exception can be of genuine service to the alert manager.

It has been said that it only takes 2 weeks to train an operator to tear up a machine, but it takes 5 years to train a mechanic to fix it. When mechanic competence is low, the shop should be organized so that most of the mechanics become specialists. Only the very best ones should be used as inspectors for diagnosing troubles, overseeing repairs, and checking work before the machine is placed in service. Less competent mechanics must be relegated to specialties such as brakes, ignition, chassis and body, and other limited functions. In that way the general competence of the shop can be raised.

Training of mechanics should never be neglected. An excellent opportunity for upgrading mechanics is provided by the manufacturer through the medium of factory training or training at special schools set up at regional offices or sometimes even in mobile trailers which can come to the work site. Not only should men be highly skilled in their own specialty, but they should also be taught another skill so they can cover for an absent mechanic or augment some shop in an emergency. Training is double-edged: it benefits not only the mechanic but also the owner. Training policy should be established and carried out with diligence.

11
Equipment Safety

BACKGROUND

Before passage in 1970 of the Williams-Steiger Act, Public Law 91-596, known as the Occupational Safety and Health Act of 1970 (OSHA), the federal government had not been deeply involved in construction safety. Many federal agencies had established safety regulations and applied them to their operations. Both the U.S. Army Corps of Engineers and the Bureau of Reclamation had written safety orders which applied to their civil works contracts.

Several states also had construction safety orders and enforcement agencies. California was one of the pioneers, with safety regulations dating back to 1914, when the Safety Department was created as a branch of the Industrial Accident Commission. In 1927 the Industrial Accident Commission was reformed into the Division of Industrial Accidents and Safety. Another reorganization of the Department of Industrial Relations, in 1945, divided that division into the Division of Industrial Safety, Division of Industrial Accidents, and the State Compensation Insurance Fund. California is one of the half dozen states now allowed by OSHA to enforce its own existing safety orders as permitted by the Act.

OSHA: THE LAW

OSHA was the landmark legislation regulating construction safety passed by the Ninety-first Congress and approved Dec. 29, 1970. It became the law of the land on Apr. 28, 1971, 120 days later. The Act applies to all employers engaged in interstate commerce (interpreted liberally) and on contracts supported by most federal appropriations such as the Highway Fund, various federal housing funds, etc. Exempted from the Act are federal agencies, and state, county, and municipal government agencies. The states are encouraged to develop their own programs and to take over the administration and enforcement of their approved regulations. A generous infusion of federal money will stimulate the states to accomplish this objective.

For contractors who have not operated under a strict state safety code, OSHA has come as quite a shock. The law now requires many things which should have been done but which the industry had never got around to doing voluntarily. While the usual direction of state codes has been abatement, OSHA requires punitive action against violators. Section 5 of Public Law 91-596 (OSHA) enjoins the employer (1) to provide each of his employees employment and a place of employment free from recognized hazards likely to cause death or serious physical harm, and (2) to comply with safety and health standards in the Act. Section 17 sets forth the penalties for the employer, some of which are:

1. Not more than $10,000 fine for each willful or repeated violation
2. Up to $1,000 fine for receiving a citation for a serious violation
3. Up to $1,000 fine for receiving a citation for a violation not of serious nature
4. Not more than $1,000 fine for each day a cited violation goes uncorrected
5. Not more than $10,000 fine or 6 months' imprisonment for a violation which results in death, except that for second and subsequent convictions the limits are $20,000 and 1 year
6. Up to $1,000 fine for violation of the posting requirements prescribed by the Act

Section 17 goes further to say that anyone may be assessed:

1. Up to $1,000 fine or imprisonment for up to 6 months for giving advance notice of an inspection
2. Up to $10,000 fine or imprisonment for up to 6 months for knowingly making a false statement, representation, or certification in any application, record, report, plan, or other document filed or required to be maintained pursuant to the Act

Although the Act admonishes each employee to comply with occupational and health standards and all rules, regulations, and orders issued pursuant to the Act which are applicable to his own actions, no penalties are assessed for these violations. Discipline is placed in the hands of the employer.

RECORD KEEPING UNDER OSHA

There are presently three important records required by OSHA. The booklet "Recordskeeping Requirements under the Williams-Steiger Occupational Safety and Health Act of 1970" may be obtained from the regional OSHA office or the Department of Labor. It describes these records as follows:

Form No. 100: Log of occupational injuries and illnesses Each occurrence must be logged within 48 hours. The person responsible for the accuracy of the log must initial each entry and change. Logs must be kept current and retained for 5 years following the end of the calendar year of the entry.

Form No. 101: Supplementary record of occupational injuries and illnesses This record must show background data for the entries in the log above. Causes and outcome must be entered.

Form No. 102: Summary of occupational injuries and illnesses Previously recorded occurrences must be summarized at the end of each year. This must be accomplished by 30 days after the year end and posted in a conspicuous and accessible location.

COST OF ACCIDENTS

The increased emphasis on safety provided by OSHA will ensure a safer environment for the worker and furnish an economic incentive for the contractor as well. Accidents cost a lot of money. The cost of accidents may be divided into two parts, direct and indirect.

EQUIPMENT SAFETY

Insurance is considered a direct cost. General insurance coverage plus workmen's compensation insurance and public liability and property damage insurance are costs regularly paid by the contractor. A safe contractor gets a direct saving in these costs in terms of reduced premiums and rebates at the end of the policy period. On a large job, this can mean a savings of thousands of dollars—sufficient incentive to maintain a good, strong safety program. Most successful and progressive contractors have done this in the past without the threat of OSHA. Now a good safety program will be mandatory for all contractors coming under the Act.

Indirect costs are numerous, some tangible and some intangible. Many are given lip service with little effort to reckon their magnitude. The following is a partial list of some well-established indirect costs:

1. Repair cost
2. Loss of revenue
3. Reduction in shop effort on scheduled maintenance
4. Overwork on remaining machines, increasing depreciation due to wear and tear
5. Rental of replacement machine while broken machine is in repair
6. Increase in parts inventory to cover additional parts replacement
7. Increase in overhead because of additional administrative responsibilities

In computing these indirect costs, all ancillary costs should be considered. Repair cost should include not only the actual cost of repair parts and labor to restore the machine to safe operating condition but also the cost of mechanics called to the scene of the accident and their transportation cost (usually a pickup truck or some emergency vehicle). Towing cost should be included as well as temporary repairs or patching to make the machine transportable.

Loss of revenue is self-evident. The generation of revenue is the sole purpose of owning a machine. When it is not able to work because of an accident or breakdown, it should be penalized by charging any revenue losses to its account. Revenue may also be lost when associated machines are not operated because the broken machine is a key machine. Thus, a deadlined shovel should be penalized for trucks which have no material to haul.

When accidents occur, repairs are usually made on an

emergency basis to get the machine back to production and thus avoid revenue loss. This becomes an unscheduled repair which interferes with scheduled maintenance in the shop. Not only mechanics' labor and supervision are stolen from the established schedule, but shop space is utilized which would have been available if no accident had occurred.

Usually when a machine is involved in an accident and rendered inoperable, an effort is made to cover its production by overworking the other machines in the group. The pressure is unconsciously placed on the other operators to make up the difference in production. An alternate solution is to work overtime with the remaining machines. In either case, additional wear and tear take place in the machines, and they depreciate more rapidly.

If a key machine is involved in an accident and work cannot proceed without it, a replacement may be acquired by rental or lease. When this is done, some of the revenue loss is alleviated, but the rental charges should be made against the down machine.

It is conceivable that frequent accidents may necessitate an increase in parts inventory. This cost, too, must be borne by machines which must be repaired because of accident damage. Where many machines are involved, increased parts inventory may entail substantial investment not otherwise required.

Finally, the increase in overhead costs will be significant because of accidents. One may tend to rationalize this by assuming that the supervisor would be there anyway. This may be true, but he could certainly be doing something more constructive toward the job than handling an accident emergency.

While everyone may not agree that all these indirect costs can be identified and are valid charges against an accident, one must admit that they accrue. The direct insurance costs are not difficult to establish because of the premiums which must be paid. It has been estimated that the ratio of indirect cost to direct cost will be 4 or 5 to 1. In the face of these costs, one cannot deny that a good safety program which will return some of this money to the contractor is a smart course of action.

ELEMENTS OF SAFETY

Safe equipment operations involve the totality of a construction job—personnel, equipment, and environment. These three elements of safety are intimately combined on the job, each affecting the other in such a way that it is sometimes difficult to separate them. By personnel is meant people, all the people involved

EQUIPMENT SAFETY

on the job. This includes everyone from top management to the common laborers. Equipment includes all the production equipment, auxiliary equipment, service equipment, and any other machines which contribute to the work. Environment means all the conditions, circumstances, and influences surrounding the job. This includes the climate, air, water, weather, and atmosphere. All these things together constitute the elements of safety.

BLS SURVEY OF CONSTRUCTION INDUSTRY

In the first survey of the construction industry by the Bureau of Labor Statistics under the new record-keeping system required by OSHA, it was reported that approximately 400 heavy-construction workers died as a result of job-related injuries or illnesses during the last 6 months of 1971. Under the new OSHA system, the basis of reporting is 100 man-years work versus the American National Standards Institute (ANSI) use of 1 million employee-hours worked, used formerly. This first survey ranked construction high on the list of recordable occupational injuries and illnesses. Of these reportable cases, 95 percent were injuries and the remaining 5 percent illnesses. The top of the list is in the following order:

Industry	*Rate per 100 man-years*
Lumber and wood products	27.6
Fabricated metal products	25.6
Primary metal industry	23.6
Water transportation	23.1
Construction	22.4
General building contractors	24.7
Heavy-construction contractors	22.1
Specialty contractors	21.1

Within the construction industry there was a variance in the rate with the size of the contractor, the highest rate being for contractors with 100 to 249 employees with a rate of 30.3 and the lowest for contractors with 1 to 19 employees at a rate of 15.4. In heavy construction there were 6.4 cases of lost workdays and 15.6 cases without lost time per 100 man-years of work. This is not greatly different from the results obtained by deStwolinski[1] in

---

[1] Lance W. deStwolinski, A Survey of the Safety Environment of the Construction Industry, *Stanford Univ. Constr. Inst. Tech. Rep.* 114, 1969.

his survey of Operating Engineers Local No. 3 in California, Nevada, and other western areas. He found a rate of lost-time accidents of 4.6 per 100 man-years of work for this group in 1969. He found that near misses, which were not reportable, occurred as follows:

Frequency	Percent of local membership
At least once a month	75
At least once a week	20
At least once a day	13

He also found a significant correlation between workers having lost-time accidents and those having minor injuries, traffic accidents, and near misses.

PERSONNEL

People are the most important ingredients in any job. Certainly they are the most influential factors in safety. Simonds and Grimaldi[1] assert that human factors play a major role in causing 85 percent of all accidents. It often has been said that accidents do not happen, they are caused. Modern thinking prefers to believe that accidents result from a series of related events following in sequence and contributing to the occurrence of the accident. Accident prevention strives to interrupt this chain of events. Looking at the records of the National Safety Council and the National Transportation Safety Board, one can surmise that out of every 20 contributing factors, one will be equipment failure and the other 19 man-related.

Probably the most important factor contributing to safety in any field is attitude. This is extremely important in equipment operations involving large, high-speed machines. What a man does with a jackhammer will usually only affect himself, but what an operator does with a self-propelled scraper with a loaded weight of 120 tons traveling 40 mi/hour may endanger many people as well as a piece of equipment worth a quarter of a million dollars. Many things affect attitude; it is a part of morale. The worker's associates, and especially his family, have a profound effect on his attitude toward working safely. A family argument

---

[1] R. H. Simonds and J. V. Grimaldi, "Safety Management," Irwin, Homewood, Ill., 1956.

EQUIPMENT SAFETY

before leaving for work in the morning may set the stage for a serious accident later in the day.

Noise, dust, and heat are deleterious to people. Since earthmoving operations are characterized by all three, people have become conditioned to them. Some relief may be forthcoming from OSHA, because the law now requires the employer to limit the exposure of the worker to these conditions. This will be discussed later under equipment requirements.

Some attention is now being focused on fatigue as a contributory factor to accidents. Long working days or 7-day work weeks bring in a lot of overtime pay to the worker but also increase his risk of accident. Heat also contributes to fatigue as well as causing some discomfort. It has become clear that as fatigue increases, production drops and accidents occur more frequently.

PERSONAL SAFETY GEAR

It is incumbent on the employer to supply the employee with safety equipment for his person. This gear will vary with the type and extent of exposure and may include any or all of the following:

Hard hat
Safety shoes
Safety goggles
Heavy gloves
Reflector vests
Ear protectors
Respirators

For his own safety, the employee should avoid loose clothing (especially loose cuffs), scarves and ties, rings, wrist watches, and anything else which might get caught in wheels, belts, or other moving parts.

EQUIPMENT

The effects of the new emphasis on safety and health through OSHA and EPA will be felt by both equipment manufacturers and users. In particular, new requirements for noise abatement, dust control, roll-over protective structures (ROPS), fenders, and a better environment for both the equipment operator and the public will pyramid costs for the years to come. And this is not all bad. Many of these amenities will not only enhance safety and health but also increase comfort and add to production. It is hoped that the latter benefits will offset the capital costs of the improvements.

REFERENCE CODES AND STANDARDS

OSHA is really a compendium of many existing codes, standards, specifications, and state regulations. A partial listing of these sources by subjects which relate specifically to equipment follows:

Subject	Source
ROPS	1,2,3,4
Fenders	1
Brakes	1
Cranes and derricks	1,5,6
Noise control	6,7
Respiratory protection	6,8,9,10

Key: 1 = Society of Automotive Engineers; 2 = U.S. Army Corps of Engineers; 3 = Bureau of Reclamation, Dept. of Interior; 4 = Division of Industrial Safety, Calif. Dept. of Industrial Relations; 5 = Construction Industry Manufacturers' Assn.; 6 = American National Standards Institute; 7 = U.S. Dept. of Labor; 8 = Bureau of Mines, Dept. of Interior; 9 = National Safety Council; 10 = Bureau of Labor Standards, Dept. of Labor.

The September 1972 issue of *Construction Methods and Equipment*[1] lists a large number of specifications with the names and addresses of sources.[2]

Although many of the OSHA requirements for equipment safety were adopted by several progressive states several years

---

[1]Safety Standards and Specs, *Constr. Methods Equip.*, pp. 62–64, September, 1972.

[2]Other useful information to supplement OSHA will be found in the following documents: U.S. Dept. of Labor, Bureau of Labor Standards, Safety and Health Regulations for Construction, *Fed. Reg.* vol. 36, no. 75, Apr. 17, 1971; State of California, Dept. of Industrial Relations, Construction Safety Orders, Office of Procurement, Documents Section, Sacramento, Calif. 1973; U.S. Army Corps of Engineers, General Safety Requirements, EM 385-1-1, Mar. 1, 1967; U.S. Dept. of the Interior, Bureau of Reclamation, Safety and Health Regulations for Construction, Sept. 1971; U.S. Dept. of Labor, Bureau of Labor Standards, Safety and Health Standards for Construction, Apr. 18, 1971–May 31, 1972; Society of Automotive Engineers, *SAE Handbook* [*New Requirements for Equipment Safety*], New York, annual.

EQUIPMENT SAFETY

ago, all represent new requirements in a majority of the states, and some will bring change to all states. It is strange indeed that the equipment manufacturer carries only an indirect burden for the installation of the new devices; the contractor (employer) is primarily responsible for compliance with the law. Only the latter can be penalized legally: the manufacturer cannot. Under the new provisions for safe equipment will be found the following requirements:

1. ROPS
2. Safety belts
3. Canopies
4. Fenders
5. Emergency, service, and parking brakes
6. Reverse alarms
7. Noise suppression
8. Dust alleviation
9. Crane boom-angle and load indicators

Most of these items are controversial. Not all of the problems have been solved in designing new equipment to satisfy the new requirements. Especially difficult and costly will be the problem of retrofitting existing machines which were not designed with such things as ROPS and load indicators in mind.

Another problem which has not been solved is that OSHA requirements for construction differ from those established by the Department of Agriculture for logging and the Bureau of Mines for mining (hence, sand and gravel pits and quarries). Where these rules differ it may not be possible fully to utilize equipment that might otherwise perform a satisfactory service.

Since the several states not only have the option but are encouraged to run their own safety programs, it may not be possible to transfer certain equipment between jobs in different states. New York, for example, requires that the design of all ROPS be approved by the Board of Standards and Appeals. As a consequence, a ROPS satisfactory in a nearby state may not be legal in New York.

ROPS

OSHA regulations for roll-over protective structures were promulgated on April 5, 1972.[1] ROPS are required on all site-clear-

---

[1] Safety and Health Standards for Construction, *U.S. Dept. Labor Occupational Safety Health Administration Pam.* OSHA 2061, and *Fed. Reg.*, vol. 37, no. 66, Apr. 15, 1972.

ing equipment and the following earth-moving equipment manufactured after Sept. 1, 1972:

Scrapers
Loaders
Wheel tractors
Crawler tractors
Bulldozers
Off-highway trucks
Graders
Agricultural and industrial tractors
Similar equipment

All equipment manufactured before the mandatory date of Sept. 1, 1972, must be retrofitted on the following schedule:

Date of manufacture	Retrofit deadline
Jan. 1, 1972–Aug. 31, 1972	Apr. 1, 1973
July 1, 1971–Dec. 31, 1971	July 1, 1973
July 1, 1970–June 30, 1971	Jan. 1, 1974
July 1, 1969–June 30, 1970	July 1, 1974
Before July 1, 1969	July 1, 1975

Off-highway trucks, compactors, and skid-steer rubber-tired tractors are held in abeyance pending promulgation of standards. Agricultural and industrial tractors of less than 20 engine horsepower are exempted. The Construction Industry Manufacturers Association has estimated that the cost of retrofitting ROPS on the 625,000 construction machines manufactured from 1969 to 1972 will involve a direct cost of over $1 billion.[1] The high cost of ROPS is indicated by the Caterpillar list price of $1,426 for a new D9 (1973) and $1,895 for a retrofit.

The design of ROPS is covered in the "Safety and Health Standards for Construction"[2] issued on Apr. 5, 1972. In general, these structures must sustain double the weight of the machine, meet a side thrust requirement, allow a vertical clearance of at

---

[1] OSHA in Construction Spells CHAOS, *Constr. Methods Equip.*, p. 78, October 1972.
[2] Ibid., pp. 14–23.

least 52 inches from the working deck for access, and be labeled as follows:

1. Manufacturer's name and address
2. ROPS model number
3. Machine make, model, or series number it is designed to fit

ROPS will be considered satisfactory if they meet the requirements of CAL/OSHA,[1] Corps of Engineers,[2] or Bureau of Reclamation.[3] The standards used as a basis for the regulation are SAE Standards J320a, J394, J395, and J396.

SEAT BELTS

Seat belts are required on all motor vehicles and earth-moving equipment on which ROPS are installed. These belts are required to meet SAE J386-1969 or SAE J333a-1970. Seat belts need not be provided for equipment designed only for standup operation.

FENDERS

Rubber-tired earth-moving equipment whose speed exceeds 15 mi/hour must be equipped with fenders meeting the requirements of SAE J321a-1970. There is a provision under section 1518.2[4] whereby the contractor may request a variance from the rules when substantial engineering or practical difficulties exist. A written application must be filed in triplicate describing the variation and reasons therefore.

BRAKES

All carth-moving equipment is required to have a braking system meeting the requirements of the following:

Equipment	Recommended practice
Self-propelled scrapers	SAE J319b-1971
Self-propelled graders	SAE J236-1971
Trucks and wagons	SAE J166-1971
Front-end loaders and dozers	SAE J237-1971

---

[1] State of California, Construction Safety Orders, op. cit.
[2] EM 385-1-1, op. cit.
[3] Safety and Health Regulations for Construction, op. cit.
[4] See *Fed. Reg.*, vol. 36, no. 75, p. 7341, Apr. 17, 1971.

Basically, the law says that the service braking system must be capable of stopping and holding the equipment fully loaded as specified in the above recommended practices. The practices specify service, parking, and emergency stopping systems. The parking system of a machine, for example, must be capable of holding the fully loaded machine on a 15 percent grade on dry swept concrete.[1] Also required is a warning system to alert the operator when the stored energy drops below 50 percent of the manufacturer's specified maximum energy operating level.

REVERSE ALARMS

All motor vehicles having an obstructed view to the rear must be equipped with an audible alarm distinguishable above the surrounding noise level or be backed only when a spotter signals it is safe to do so. The same requirement is imposed on all earth-moving or compacting equipment. Additionally, all bidirectional machines such as rollers, compacters, front-end loaders, bulldozers, and similar equipment must have a horn, distinguishable from the surrounding noise level, which must be operated as needed when the machine is moving in either direction. Reverse alarm devices are described in SAE J994.

NOISE SUPPRESSION

This requirement stems from the OSHA regulations which are designed to protect the equipment operator and the EPA requirement to protect the public from loud noise. These rules may on occasion be in conflict. Here the OSHA requirements will be set forth. Article 1518.52 of the Safety and Health Regulations for Construction[2] covers the noise limitations for employees. Table D-2 of this article is summarized in Fig. 11-1. An equation is also given for compound noise exposure, with explanation, in Article 1926.52 of Safety and Health Standards for Construction.[3] In this equation, the exposure exceeds permissible levels when F_e is greater than 1.0:

$$F_e = \frac{T_1}{L_1} + \frac{T_2}{L_2} + \ldots + \frac{T_n}{L_n}$$

where F_e = equivalent noise-exposure factor
T = period of noise at constant level (see Fig. 11-1)

---

[1] See CAL/OSHA, op. cit., art. 1591.
[2] Op. cit., p. 7348.
[3] Op. cit., p. 8.

EQUIPMENT SAFETY

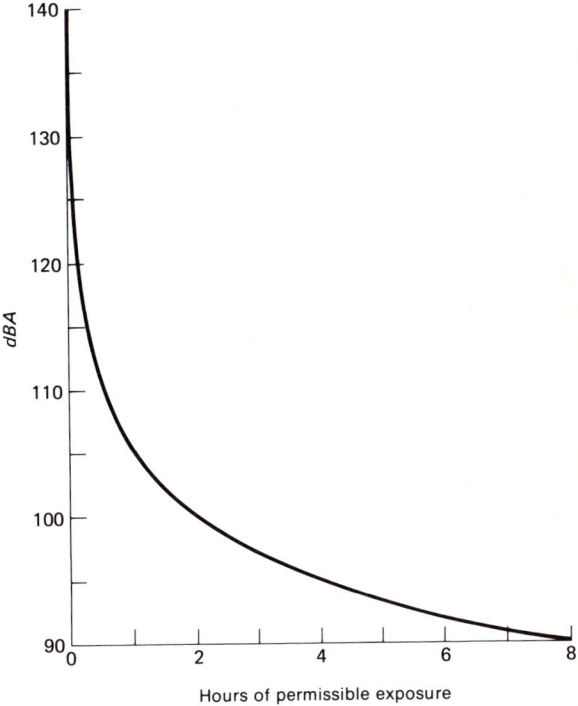

Fig. 11-1 Permissible noise exposures.

When employees are subjected to unacceptable noise levels, feasible administrative or engineering controls shall be utilized. Where it is not feasible to reduce noise levels by these means, ear-protective devices shall be provided and used.

DUST ALLEVIATION

Dust is not only a problem of respiration but also one of eye protection. Articles 1518.55, 1518.102, and 1518.103 of the Regulations[1] require respirators and protective goggles when dust levels exceed the Threshold Limit Values of Airborne Contaminants for 1970.[2] Adequate ventilation in tunnels and closed spaces, wet drilling, and watering of roads and excavation areas

---

[1] Ibid , p. 7349.
[2] American Conference of Governmental Industrial Hygienists; see Safety Standards and Specs, loc. cit.

are the usual means of control. If these measures are not adequate, personal protective devices must be furnished and used.

CRANE INDICATORS

The OSHA regulations require that all cranes be equipped with a boom-angle indicator and a load-indicating device in good working order. A load-moment device will be accepted in lieu of them.

ENVIRONMENTAL CONTROLS

One of the best ways to alleviate some of the unsafe and unhealthy conditions associated with equipment operations is to control the operator's environment. This can be done by designing protective cabs for all feasible equipment. The cab should be built with the ROPS, air-conditioned, and maintained with a positive internal pressure to exclude all dust and other possible contaminants. This would not only alleviate the noise and dust problems but would provide the operator with air-conditioned comfort to reduce heat fatigue or exposure to cold. Production would be enhanced and the additional costs for the cab offset by the increased production. SAE J169 gives the design guidelines for air-conditioning systems for construction and industrial equipment cabs.

POLICY DEVELOPMENT

The development of policy related to safety of equipment operations is no longer optional anywhere in this country. Federal regulations[1] state that it is the responsibility of the employer to initiate and maintain such programs as may be necessary to comply with OSHA. Frequent and regular inspections of the jobsites, materials, and equipment are to be made by competent persons. The use of unsafe machinery, tools, materials, or equipment is prohibited. An important new rule is the requirement that employers permit only those employees qualified by training or experience to operate equipment and machinery. This implies not only a training program but also licenses or permits for qualified operators. Another important new rule[2] requires that all roadway grades for the use of construction equipment be designed by a qualified engineer, competent in this field, and constructed to accommodate all the equipment using the road.

---

[1]Safety and Health Regulations for Construction, art. 1518.20.
[2]Ibid., art. 1518.602 (a)(3).

EQUIPMENT SAFETY

WRITTEN POLICY STATEMENT

The commitment of the contractor to a policy of safe operations should be stated in writing. The statement should be disseminated to all employees and included in the front of the contractor's safety manual or safety regulations pertaining to the job. This same policy statement should be posted on all jobs as well, so that the employees will understand it as a continuing commitment on the part of the employer. This policy statement should also be read to all new employees in their orientation, so that they will understand from the beginning that safety is the order of the day.

ESTABLISH GOALS

These goals should be established with a full awareness of all the laws which relate to safety: municipal, county, state, and federal. Usually compliance with the toughest law is required, so it may not be sufficient to consider only the federal law. Once these laws have been sifted out to determine just what the contractor must comply with, and this may take some wringing out by the courts, programs should be formulated by the contractor to bring his operations into compliance with the laws.

    These programs are going to be necessary for most contractors until manufactured equipment and practices are at least brought into compliance with OSHA. While state and local laws used to be lax, especially in remote areas, the new federal act is applicable everywhere. The fact that only a half dozen states are ready to take over enforcement from OSHA indicates that many of them have a long way to go in raising their standards. Retrofitting programs will be required to comply now with ROPS and fender requirements. Later, when the rules have been clarified, noise controls and emission controls will also be retrofitted. The costs of these two may well exceed the cost of retrofitting ROPS.

    A program will also be required for training and licensing equipment operators. This has long been overdue for many contractors. While the apprenticeship program has endeavored to train operators in the past, the burden is now on the contractor for screening and perhaps augmenting the training already provided. At least, it is now up to the employer to be sure that his operators are qualified before they are assigned to work.

LIMITATIONS OF SAFETY PROGRAMS

The law, like most codes, really only stipulates the minimum standards for safety. How far the contractor wishes to go beyond the minimum required by the law will be determined by his ability to curb costs by operating safely and not accepting the high costs of accidents. In calculating these costs, he must consider all the costs mentioned in an earlier section. It is the resolution of all insurance (direct costs) and indirect costs which will tell him how much he can expect to recover through risk management. His savings in direct and indirect costs should at least exceed the cost of his safety program.

EMPLOYEE EDUCATION

Employee education must begin at the top and work downward. Managers must be motivated to pursue a program in which safety is given priority over production. Next, supervisors down through the level of foremen must be convinced that an adequate and active safety program will result in increased returns to everyone. Finally, employees must be made to understand that only a safe job is acceptable. The education of the employee is a part of the training program which should be implemented with vigor. The best means of education for safety is by example. Unsafe acts or equipment must not be tolerated by any level in the organization except for emergent and compelling reasons.

ORGANIZING FOR SAFETY

Once management has become dedicated to safe operations, the establishment of a program is undertaken in three steps: (1) determine the effort to go into the program, (2) establish a program which reflects the effort to go into it, and (3) set up an organization to carry it out.

The amount of effort to go into a program will begin with the legal requirement; this will be the minimum. After that, the contractor must set his own limits based on his assessment of costs and savings, his own personal views on safety and human relations, and the image he wishes to project both to his employees and to the public at large. An analysis of past accidents should be made to assist in determining the direction and effort to be expended in the safety program. The contractor's own training and experience also will have great influence in this decision. For humanitarian and economic reasons alone, he should be propelled in the direction of safety.

The program established will reflect the amount of effort and in which directions the program should go. Here, the nature of the work will have considerable weight. It would be useless for a constructor to have elaborate instructions and training in blasting and handling explosives unless he is in a business where explosives are required. A building contractor would not need this emphasis, for example, while a tunneling contractor would find them imperative.

Now that the dimensions of the program have been determined and the program established, an organizational structure is required to implement it. It has often been said that safety and morale are primary responsibilities of the man in charge of the job. The chief burden of responsibility of running a safe job belongs with the project manager or the top man on the job in the field, whatever he is called. Under him should be a staff man with primary responsibility for safety. On a large job the safety man should spend full time in this function. On a small job, part time only will suffice. He should advise and report directly to the man in charge. He should act as liaison with other agencies or activities in safety matters; but he is still acting in a staff capacity.

Safety functions should be routed directly through the line chain of command. Orders and advisories related to safety can be prepared by the safety officer but should always be disseminated downward from the project manager to superintendents, foremen, and ultimately the workers. Whatever line and staff responsibilities exist for safety, the project manager should always be aware that the responsibility for running a safe job is his even though he may share it in part.

ESTABLISHMENT OF POLICY

Once the step-by-step process has been carried out in establishing goals, setting priorities, and organizing for safety, the policy should be clearly stated by writing it down. The three most important parts of policy are scope, authority and responsibility, and the organization for and administration of the policy.

The scope of the policy will depend on the effort to be expended and whether or not the contractor wants to extend it to cover some aspects of safety off the job as well as on. Safety off the job has an effect on absenteeism and ability of the worker to do his normal work. Safe automobile-driving habits may have more to do with a man's presence or absence on the job than some job-related aspects of safety. In remote areas, poisonous snakes, vicious animals, or other local hazards on and off the job may become a part of a program. At least the requirements of the federal regulations outlined in OSHA[1] must be met. This compli-

---

[1] Safety and Health Regulations for Construction, Sec. 1518.20.

ance must include programs to comply with the law, frequent and regular safety inspections, a prohibition of the use of unsafe tools and equipment, and authorization of only competent and qualified employees to operate machinery and equipment.

Authority and responsibility go hand in hand. They should be delegated to all levels of supervision in order to implement the established safety program. Each higher level should include more authority and responsibility than the levels below. Who, under what circumstances has the right to decide to violate safety rules should be carefully weighed and clearly enunciated.

Finally the organization for carrying out the policy and the administrative procedures by which it will be accomplished should be stated. The determination of who will make inspections and who will report violations should be a part of the administration of the program. What will be done with violation reports? Who will discipline offenders? What will the punishment be? These are questions which should be answered. In order to comply with OSHA, certain forms must be filled out, retained, or posted. Who will do this?

Accident reports should be reviewed in order to learn from them and correct unsafe practices or unsafe equipment. The procedures for this review should be outlined. Also to be stated is how the results of these reviews are to be brought to the attention of everyone in the organization, especially those likely to be affected by them.

As a completion of the administration of the program, the matter of amending the safety rules should be set forth. When the review of an accident shows that a new rule should be established or an existing rule changed, the procedure for doing this should be enunciated. This is the final response to a well-conceived system of safety management.

TRAINING FOR SAFETY

This begins when the first man is hired on the job and never stops until the job is finished. The first thing to be done is to introduce the new employee to his job and its surrounding conditions. He should be shown or read the management statement on safety policy. He should be instructed on steps to be taken in case of an accident to himself or a fellow worker. He should be taken to danger areas on the site, and it should be explained why they are dangerous. Any health hazards in the production operation or job site should be explained thoroughly, especially how to avoid them or protect himself. He should be properly schooled in the operation of equipment he may be expected to operate. He should be properly tested and licensed for this purpose when his qualifications so indicate. And finally, he should be given a company booklet containing information about job safety and how he relates to the safety program.

Work assignment should follow management's concern for the safety of the new employee. If he is not knowledgeable about the safe accomplishment of the work expected of him, he should be instructed. He should also be advised of any need for personal protective equipment such as hard hat, gloves, ear protection, respirator, or goggles, and any or all of these items should be provided for him as needed.

Toolbox or tail-gate safety meetings should be conducted by supervisory personnel once a week. The California Safety Code requires them to be held at least every 10 working days on the job or as needed to emphasize safety. These meetings normally should be kept short, no more than 10 min. There should be a dialogue between the supervisor and the workers. Recent accidents, new hazards, recurring unsafe acts, and other matters of current importance should be discussed. Monthly (or more frequent if necessary) meetings of all foremen with management should be held to discuss safety problems and accidents that have occurred.

Safety aids will serve a useful purpose in the training program. They may be obtained from many sources, among which are trade associations, technical societies, government agencies, and insurance companies. Some of the organizations which disseminate useful aids are:

National Safety Council
American National Standards Institute
Associated General Contractors
American Institute of Steel Construction
National Asphalt Pavement Association
American Society of Testing Materials
Society of Automotive Engineers
Construction Industry Manufacturers Association

DIRECTION OF EQUIPMENT SAFETY

While the impact of OSHA and its requirements for ROPS, fenders, and other changes in equipment is beginning to be felt, the future will bring even more changes. Heat, dust, and noise are all good targets for corrective action; OSHA and the EPA have the ammunition.

The air-conditioned, environmentally controlled cab combined with ROPS is a good goal to which to aspire. Heat, dust, and noise are discussed in detail by deStwolinski[1] in his excellent technical report. The most satisfactory solution to these imposing problems appears to be the enclosed cab. Such equipment is now available as an option from several manufacturers. An approximate cost of these items for a Caterpillar D9 tractor, installed at the Peoria, Ill., factory and delivered in Oakland, Calif. (exclusive of state sales tax), was in 1973:

Cab with ROPS	$3,190
Air conditioning	2,130
Acoustic insulation	1,000
Freight	120
Total	$6,440

It has been estimated by CIMA that the retrofitting cost of noise and emission controls will far exceed that of ROPS. There are already indications of what the penalties may be in increased operating cost and impaired performance.

One item affecting safety which has received almost no publicity is the standardization of controls. The diversity of controls on almost all construction equipment is greater than manual gearshifts in automobiles before the standard shift. One has to go back 40 years to recall that Buick, Dodge, and Chevrolet all had different gearshift patterns. It was easy for a confused driver to put the car in reverse gear rather than forward. Today, there are cranes and excavators with reversed controls. This is dangerous. There should be some intensive thought in the direction of standardizing these control mechanisms, especially on mobile equipment such as tractors, loaders, cranes, excavators, graders, trucks, and scrapers.

More automated earth-moving machines in the future will provide less exposure of people to safety hazards. This is especially true of night operations where field lighting is less than adequate. The time will come when it is possible for one operator to control several machines or perhaps even an earth-moving system. The end result will be less risk to the person involved.

---

[1]Lance W. deStwolinski, Occupational Health in the Construction Industry, *Stanford Univ. Constr. Inst. Tech. Rep.* 105, 1969.

EQUIPMENT SAFETY

A new trend is evidenced by the California State Department of Industrial Safety, Division of Industrial Safety, to provide a full range of consultative services to employer and employee groups. OSHA has been criticized because when an inspector arrives on a job and finds a safety violation, he must take action. With the state safety consultant, he may be called to inspect the site for council and advice without issuing a warning or citation if something is wrong. This should encourage employers to seek the help of the professionals in code enforcement without fear of punitive action.

The truth is that the safety environment looks better in construction than it has in the past. Twenty years ago it was almost ignored: risks were accepted as part of a construction career. Now, the industry is awakening. A safer job is a better job for both employer and employee. The net result is a better life for all.

appendix 1
List of Variables

A_0 = present age of present machine for depreciation
A_1 = present annual expected maintenance and operating costs of present machine
A_2 = annual expected maintenance and operating costs of new item if purchased now
B_1 = difference between initial cost A_1 and upper limit of cost as present machine ages
B_2 = difference between initial cost A_2 and upper limit of cost as new machine ages
C_1 = present salvage value of present machine
C_2 = purchase price of new machine presently available
C_3 = coefficient to determine rate of decline in salvage value of present machine
C_4 = coefficient to determine rate of decline in salvage value of future machines
C_5 = present cost of discrete, redundant expense
C_6 = coefficient of decline of discrete expense between replacements
C_7 = book value of present machine

LIST OF VARIABLES

D_1	= present worth of double-declining-balance depreciation on present machine
D_2	= present worth of double-declining-balance depreciation on future machines
D_3	= coefficient to determine rate of allowable depreciation for present machine
D_4	= coefficient to determine rate of allowable depreciation on future machines
D_5	= present worth of straight-line depreciation on present machine
D_6	= present worth of straight-line depreciation on future machines
exp	= base of natural logarithms = 2.71828+
E_0	= present worth of all expected costs
E_1, E_2	= present worth of partial expected maintenance and operating costs of present machine
E_3, E_4, E_5	= present worth of partial expected maintenance and operating costs of future machines
E_6	= present worth of capital costs of present machine
E_7	= present worth of capital costs of future machines
E_8	= present worth of taxes on present machine
E_9	= present worth of taxes on future machines
E_{10}	= present worth of all future discrete expenses on present machine
E_{11}	= present worth of all future discrete expenses on future machines
F	= ratio of amount of financing to total cost of new machine
G	= coefficient to determine rate of approaching upper limit of revenue (increase in productivity)
G_1	= present worth of gain on sale of present machine
G_2	= present worth of gain on sale of future machines
I_0	= coefficient to determine rate of inflation of currency
I_1	= present worth of interest charges on loan to purchase present machine
I_2	= present worth of interest charges on loan to purchase future machines
IC_1	= present worth of investment credit when present machine is replaced
IC_2	= present worth of investment credit for replacement of all future machines
j	= number of replacement ($j = 0$ is the beginning of the first replacement)

K	= fraction of C_2 remaining after installation costs
L	= life of replacement equipment, years
M_1	= number of overhaul (first = 0) on present machine
M_2	= number of overhaul (first = 0) on future machines
N	= number of years of life remaining in present machine
P_1	= annual revenue of present machine at the present time
P_2	= annual revenue of new machine if purchased now
PIC	= percentage of qualified investment taken as tax credit
Q	= difference between initial revenue P_2 and upper limit of revenue of all future machines
R_0	= present worth of total revenue
R_1	= present worth of revenue from present machine
R_2, R_3	= present worth of partial revenue from future machines
R_4	= annual rate of simple interest on new-equipment loan, paid monthly during life of loan
R_5	= effective income-tax rate (federal and state)
R_6	= force of interest used for continuous compounding
S_1	= coefficient to determine rate of decline of annual revenue of present machine due to aging
S_2	= coefficient to determine rate of decline of annual revenue of future machines due to aging
T_1	= remaining term of loan in years on present machine
T_2	= term of loan in years on future machine
T_3	= interval in years between discrete expenses
T_4	= time in years to next discrete expense, present machine
U	= coefficient which reduces B_2 because of technological improvements in future machines
V	= coefficient of cost growth in future machine prices due to technological improvements
W_1	= coefficient to determine rate at which maintenance and operating costs of present machine approach upper limit
W_2	= coefficient to determine rate at which maintenance and operating costs of future machines approach upper limit
Z	= coefficient to determine rate of decline in maintenance and operating costs of replacement machines due to obsolescence

appendix 2
Summary of Equations

$$\text{Maximize: } R_0 - E_0 \tag{1}$$

$$R_0 = R_1 + R_2 + R_3 \tag{2}$$

$$E_0 = E_1 + E_2 + E_3 + E_4 + E_5 + E_6 + E_7 + E_8 \\ + E_9 + E_{10} + E_{11} + I_1 + I_2 \tag{3}$$

$$R_1 = P_1 \frac{1 - \exp\left[-(S_1 + R_6)N\right]}{S_1 + R_6} \tag{4}$$

$$R_2 = \frac{P_2 + Q}{1 - \exp(-R_6 L)} \frac{1 - \exp\left[-(S_2 + R_6)L\right]}{S_2 + R_6} \exp(-R_6 N) \tag{5}$$

$$R_3 = \frac{Q \exp[-GN]}{1 - \exp[-(G + R_6)L]} \\ \frac{1 - \exp[-(S_2 + R_6)L]}{S_2 + R_6} \exp(-R_6 N) \tag{6}$$

$$E_1 = (A_1 + B_1) \frac{1 - \exp(-R_6 N)}{R_6} \tag{7}$$

$$E_2 = -B_1 \frac{1 - \exp[-(W_1 + R_6)N]}{W_1 + R_6} \tag{8}$$

$$E_3 = A_2 \frac{\exp\left[-(Z+R_6)N\right]}{R_6} \frac{1-\exp(-R_6 L)}{1-\exp\left[-(Z+R_6)L\right]} \tag{9}$$

$$E_4 = B_2 \frac{\exp(-R_6 N)}{R_6} \frac{1-\exp(-R_6 L)}{1-U^L \exp(-R_6 L)} \tag{10}$$

$$E_5 = -B_2 \frac{\exp(-R_6 N)}{R_6 + W_2} \frac{1-\exp\left[-(W_2+R_6)L\right]}{1-U^L \exp(-R_6 L)} \tag{11}$$

$$E_6 = C_1\{1 - \exp\left[-(C_3+R_6)N\right]\} \tag{12}$$

$$E_7 = C_2 \exp\left[-(R_6-V)N\right]$$
$$\left\{1 + \{1 - K\exp\left[-(C_4+V)L\right]\} \frac{\exp\left[-(R_6-V)L\right]}{1-\exp\left[-(R_6-V)L\right]}\right\} \tag{13}$$

$$E_8 = R_5[(R_1+G_1) - (E_1+E_2+E_{10}+D_1+I_1)] \tag{14}$$

$$E_9 = R_5[(R_2+R_3+G_2) \\ - (E_3+E_4+E_5+E_{11}+D_2+I_2)] - \text{IC}_1 - \text{IC}_2 \tag{15}$$

$$E_{10} = C_5 \exp(-R_6 T_4) \sum_{M_1=0}^{M_1<(N-T_4)/T_3} \exp(-M_1 R_6 T_3)$$
$$\text{where } N > T_4 \tag{16}$$

$$E_{11} = \frac{C_5 \exp\{-[C_6 N + R_6(N+T_3)]\}}{1-\exp\left[-(C_6+R_6)L\right]}$$
$$\sum_{M_2=0}^{M_2<(L/T_3)-1} \exp(-M_2 R_6 T_3) \quad \text{where } L > T_3 \tag{17}$$

$$E_{10} = 0 \quad \text{where } N \leq T_4 \tag{18}$$

$$E_{11} = 0 \quad \text{where } L \leq T_3 \tag{19}$$

$$D_1 = \frac{C_7 D_3}{D_3 + R_6 + I_0}\{1 - \exp\left[-(D_3+R_6+I_0)N\right]\} \tag{20}$$

$$D_2 = \frac{C_2 D_4 \exp\left[-(R_6-V)N\right]}{D_4+R_6+I_0} \frac{1-\exp\left[-(D_4+R_6+I_0)L\right]}{1-\exp\left[-(R_6-V)L\right]} \tag{21}$$

$$D_3 = \ln \frac{N+A_0}{N+A_0-2} \tag{22}$$

$$D_4 = \ln \frac{L}{L-2} \tag{23}$$

$$D_5 = \frac{C_7}{N+A_0} \frac{1-\exp\left[-(R_6+I_0)N\right]}{R_6+I_0}\{1-\exp\left[(I_0-C_3)N\right]\} \tag{24}$$

$$D_6 = \frac{C_2}{L} \frac{\exp\left[-(R_6-V)N\right]}{R_6+I_0} \frac{1-\exp\left[-(R_6+I_0)L\right]}{1-\exp\left[-(R_6-V)L\right]}$$
$$\{1 - K\exp\left[(I_0-C_4)L\right]\} \tag{25}$$

$$I_1 = \frac{R_4 F C_2}{R_6+I_0}\{1-\exp\left[-(R_6+I_0)T_1\right]\} \quad \text{where } N < T_1 \tag{26}$$

SUMMARY OF EQUATIONS

$$I_2 = \frac{R_4 F C_2}{R_6 + I_0} \exp\left[-(R_6 - V)N\right]$$
$$\frac{1 - \exp\left[-(R_6 + I_0)T_2\right]}{1 - \exp\left[+(R_6 - V)L\right]} \quad \text{where } L > T_2 \quad (27)$$

$$I_1 \ [N \leq T_1] = I_1 \ [T_1 = N] \quad \text{when } N \leq T_1 \quad (28)$$

$$I_2 \ [L \leq T_2] = I_2 \ [T_2 = L] \quad \text{when } L \leq T_2 \quad (29)$$

$$\text{IC}_1 = \text{PIC}\{C_2 \exp\left[-(R_6 - V)N\right] - C_1 \exp\left[-(C_3 + R_6)N\right]\} \quad (30)$$

$$\text{IC}_2 = \text{PIC}\frac{C_2 \exp\left[-(R_6 - V)N\right]}{\exp\left[(R_6 - V)L\right] - 1}\{1 - K \exp\left[-(C_4 + V)L\right]\} \quad (31)$$

$$G_1 = C_1 \exp\left[(I_0 - C_3)N\right] - C_7 \exp(-D_3 N) \quad (32)$$

$$G_2 = \frac{C_2 \exp\left[-(R_6 - V)N\right]}{1 - \exp\left[-(R_6 - V)L\right]}$$
$$\{K \exp\left[-(C_4 + R_6)L\right] - \exp\left[-(D_4 + R_6 + I_0)L\right]\} \quad (33)$$

appendix 3
Table of Factors

TABLE OF FACTORS

ALL FACTORS FOR N=0.0 ARE COMPUTED TO HAVE THE FOLLOWING VALUES AT 10% RATE OF RETURN

N = 0.0

FACTOR			VALUE	FACTOR	G	S2	LIFE	VALUE
R1			0.00000	F3	0.020	0.020	2.0	8.67228
							3.0	8.67227
							4.0	8.67227
							5.0	8.67228
							6.0	8.67227
FACTOR	S2	LIFE	VALUE				7.0	8.67227
							8.0	8.67227
F2	0.020	2.0	10.29154				9.0	8.67228
		3.0	10.19806				10.0	8.67227
		4.0	10.10894				15.0	8.67227
		5.0	10.02406					
		6.0	9.94328			0.040	2.0	8.50765
		7.0	9.86648				3.0	8.43177
		8.0	9.79353				4.0	8.36000
		9.0	9.72429				5.0	8.29223
		10.0	9.65864				6.0	8.22830
		15.0	9.37971				7.0	8.16809
							8.0	8.11146
	0.040	2.0	10.09617				9.0	8.05826
		3.0	9.91524				10.0	8.00836
		4.0	9.74494				15.0	7.80337
		5.0	9.58477					
		6.0	9.43424			0.060	2.0	8.34726
		7.0	9.29287				3.0	8.20037
		8.0	9.16020				4.0	8.06321
		9.0	9.03579				5.0	7.93529
		10.0	8.91922				6.0	7.81612
		15.0	8.43993				7.0	7.70524
							8.0	7.60220
	0.060	2.0	9.90583				9.0	7.50655
		3.0	9.64313				10.0	7.41785
		4.0	9.39898				15.0	7.06501
		5.0	9.17220					
		6.0	8.96165			0.080	2.0	8.19098
		7.0	8.76629				3.0	7.97769
		8.0	8.58510				4.0	7.78103
		9.0	8.41715				5.0	7.59987
		10.0	8.26155				6.0	7.43313
		15.0	7.64133				7.0	7.27984
							8.0	7.13902
	0.080	2.0	9.72037				9.0	7.00979
		3.0	9.38127				10.0	6.89130
		4.0	9.07005				15.0	6.43392
		5.0	8.78449					
		6.0	8.52253				2.0	8.03868
		7.0	8.28230				3.0	7.76335
		8.0	8.06203				4.0	7.51263
		9.0	7.86013			0.100	5.0	7.28446
		10.0	7.67510				6.0	7.07695
		15.0	6.95876				7.0	6.88836
							8.0	6.71707
	0.100	2.0	9.53964				9.0	6.56159
		3.0	9.12922				10.0	6.42056
		4.0	8.75718				15.0	5.89139
		5.0	8.41992					
		6.0	8.11415				2.0	7.53345
		7.0	7.83692				3.0	7.60124
		8.0	7.58553				4.0	7.66649
		9.0	7.35756		0.040	0.020	5.0	7.72916
		10.0	7.15082				6.0	7.78921
		15.0	6.37197				7.0	7.84662
							8.0	7.90141
							9.0	7.95356
							10.0	8.00312
							15.0	8.21336
							2.0	7.39044
							3.0	7.39044
							4.0	7.39044
						0.040	5.0	7.39044
							6.0	7.39044
							7.0	7.39044
							8.0	7.39044
							9.0	7.39044
							10.0	7.39044
							15.0	7.39043

CONSTRUCTION EQUIPMENT POLICY

$N = 0.0$

G	S2	FACTOR F3 LIFE	VALUE	G	S2	LIFE	VALUE
0.040	0.060	2.0	7.25111	0.060	0.100	2.0	6.20072
		3.0	7.18762			3.0	6.09560
		4.0	7.12807			4.0	5.99907
		5.0	7.07232			5.0	5.91065
		6.0	7.02024			6.0	5.82982
		7.0	6.97166			7.0	5.75612
		8.0	6.92645			8.0	5.68907
		9.0	6.88444			9.0	5.62820
		10.0	6.84549			10.0	5.57308
		15.0	6.69115			15.0	5.36915
	0.080	2.0	7.11535	0.080	0.020	2.0	6.03936
		3.0	6.99244			3.0	6.20082
		4.0	6.87861			4.0	6.35754
		5.0	6.77337			5.0	6.50909
		6.0	6.67624			6.0	6.65510
		7.0	6.58675			7.0	6.79524
		8.0	6.50444			8.0	6.92928
		9.0	6.42885			9.0	7.05702
		10.0	6.35957			10.0	7.17836
		15.0	6.09345			15.0	7.68866
	0.100	2.0	6.98306		0.040	2.0	5.92471
		3.0	6.80457			3.0	6.02886
		4.0	6.64134			4.0	6.12862
		5.0	6.49227			5.0	6.22384
		6.0	6.35633			6.0	6.31439
		7.0	6.23254			7.0	6.40018
		8.0	6.12000			8.0	6.48117
		9.0	6.01780			9.0	6.55737
		10.0	5.92515			10.0	6.62882
		15.0	5.57963			15.0	6.91830
0.060	0.020	2.0	6.68944		0.060	2.0	5.81301
		3.0	6.80926			3.0	5.86340
		4.0	6.92509			4.0	5.91105
		5.0	7.03673			5.0	5.95594
		6.0	7.14401			6.0	5.99809
		7.0	7.24681			7.0	6.03751
		8.0	7.34505			8.0	6.07427
		9.0	7.43864			9.0	6.10841
		10.0	7.52758			10.0	6.14003
		15.0	7.90353			15.0	6.26369
	0.040	2.0	6.56245		0.080	2.0	5.70418
		3.0	6.62042			3.0	5.70418
		4.0	6.67573			4.0	5.70418
		5.0	6.72835			5.0	5.70418
		6.0	6.77827			6.0	5.70418
		7.0	6.82550			7.0	5.70418
		8.0	6.87006			8.0	5.70418
		9.0	6.91197			9.0	5.70418
		10.0	6.95130			10.0	5.70418
		15.0	7.11165			15.0	5.70418
	0.060	2.0	6.43874		0.100	2.0	5.59812
		3.0	6.43873			3.0	5.55093
		4.0	6.43874			4.0	5.50742
		5.0	6.43873			5.0	5.46745
		6.0	6.43874			6.0	5.43085
		7.0	6.43874			7.0	5.39743
		8.0	6.43874			8.0	5.36704
		9.0	6.43874			9.0	5.33946
		10.0	6.43874			10.0	5.31453
		15.0	6.43874			15.0	5.22318
	0.080	2.0	6.31819				
		3.0	6.26389				
		4.0	6.21340				
		5.0	6.16657				
		6.0	6.12324				
		7.0	6.08325				
		8.0	6.04644				
		9.0	6.01264				
		10.0	5.98168				
		15.0	5.86359				

FACTOR	VALUE
E1	0.00000
E2	0.00000

TABLE OF FACTORS

$$N = 0.0$$

FACTOR		LIFE	VALUE	W2				LIFE	VALUE
E3		2.0	9.58800	0.080				2.0	8.17819
		3.0	9.63205					3.0	7.96597
		4.0	9.67469					4.0	7.77028
		5.0	9.71592					5.0	7.59000
		6.0	9.75574					6.0	7.42409
		7.0	9.79414					7.0	7.27153
		8.0	9.83113					8.0	7.13139
		9.0	9.86671					9.0	7.00278
		10.0	9.90090					10.0	6.88486
		15.0	10.05142					15.0	6.42971
				0.100				2.0	8.02614
								3.0	7.75194
FACTOR		LIFE	VALUE					4.0	7.50225
E4		2.0	8.82747					5.0	7.27500
		3.0	8.90920					6.0	7.06834
		4.0	8.98853					7.0	6.88050
		5.0	9.06540					8.0	6.70989
		6.0	9.13978					9.0	6.55503
		7.0	9.21163					10.0	6.41457
		8.0	9.28093					15.0	5.88754
		9.0	9.34765	0.120				2.0	7.87796
		10.0	9.41180					3.0	7.54590
		15.0	9.69440					4.0	7.24721
								5.0	6.97863
								6.0	6.73719
								7.0	6.52022
								8.0	6.32529
								9.0	6.15022
FACTOR	W2	LIFE	VALUE					10.0	5.99305
E5	0.020	2.0	8.65874					15.0	5.41876
		3.0	8.65954						
		4.0	8.66030						
		5.0	8.66103						
		6.0	8.66172	FACTOR					VALUE
		7.0	8.66238	E6					0.00000
		8.0	8.66301						
		9.0	8.66361						
		10.0	8.66418						
		15.0	8.66660						
	0.040	2.0	8.49437	FACTOR	K	V	C4	LIFE	VALUE
		3.0	8.41938	E7	0.700	0.010	0.080	2.0	3.23238
		4.0	8.34846					3.0	2.59649
		5.0	8.28146					4.0	2.25802
		6.0	8.21828					5.0	2.04079
		7.0	8.15877					6.0	1.88582
		8.0	8.10280					7.0	1.76770
		9.0	8.05021					8.0	1.67357
		10.0	8.00089					9.0	1.59619
		15.0	7.79826					10.0	1.53114
								15.0	1.31538
	0.060	2.0	8.33423				0.100	2.0	3.35561
		3.0	8.18832					3.0	2.70318
		4.0	8.05208					4.0	2.35035
		5.0	7.92499					5.0	2.12063
		6.0	7.80661					6.0	1.95484
		7.0	7.69645					7.0	1.82732
		8.0	7.59408					8.0	1.72504
		9.0	7.49904					9.0	1.64060
		10.0	7.41093					10.0	1.56944
		15.0	7.06039					15.0	1.33350
							0.120	2.0	3.47401
								3.0	2.80366
								4.0	2.43558
								5.0	2.19288
								6.0	2.01605
								7.0	1.87914
								8.0	1.76890
								9.0	1.67770
								10.0	1.60080
								15.0	1.34693

CONSTRUCTION EQUIPMENT POLICY

$N = 0.0$

V	FACTOR E7 C4	K 0.700 LIFE	VALUE	K	V	C4	LIFE	VALUE
0.010	0.140	2.0	3.58776	0.700	0.030	0.080	2.0	4.14070
		3.0	2.89829				3.0	3.29509
		4.0	2.51425				4.0	2.83954
		5.0	2.25825				5.0	2.54368
		6.0	2.07033				6.0	2.33034
		7.0	1.92420				7.0	2.16614
		8.0	1.80627				8.0	2.03417
		9.0	1.70869				9.0	1.92486
		10.0	1.62648				10.0	1.83234
		15.0	1.35687				15.0	1.52032
	0.160	2.0	3.69706			0.100	2.0	4.29856
		3.0	2.98741				3.0	3.43049
		4.0	2.58688				4.0	2.95564
		5.0	2.31740				5.0	2.64320
		6.0	2.11848				6.0	2.41562
		7.0	1.96337				7.0	2.23919
		8.0	1.83812				8.0	2.09673
		9.0	1.73457				9.0	1.97842
		10.0	1.64750				10.0	1.87817
		15.0	1.36424				15.0	1.54127
0.020	0.080	2.0	3.62612			0.120	2.0	4.45024
		3.0	2.89921				3.0	3.55801
		4.0	2.50989				4.0	3.06282
		5.0	2.25846				5.0	2.73325
		6.0	2.07808				6.0	2.49126
		7.0	1.93987				7.0	2.30270
		8.0	1.82922				8.0	2.15004
		9.0	1.73789				9.0	2.02316
		10.0	1.66082				10.0	1.91570
		15.0	1.40285				15.0	1.55678
	0.100	2.0	3.76437			0.140	2.0	4.59596
		3.0	3.01835				3.0	3.67810
		4.0	2.61251				4.0	3.16175
		5.0	2.34682				5.0	2.81474
		6.0	2.15413				6.0	2.55834
		7.0	2.00529				7.0	2.35792
		8.0	1.88548				8.0	2.19547
		9.0	1.78625				9.0	2.06052
		10.0	1.70237				10.0	1.94643
		15.0	1.42218				15.0	1.56828
	0.120	2.0	3.89719			0.160	2.0	4.73597
		3.0	3.13055				3.0	3.79120
		4.0	2.70724				4.0	3.25308
		5.0	2.42677				5.0	2.88846
		6.0	2.22158				6.0	2.61784
		7.0	2.06217				7.0	2.40592
		8.0	1.93342				8.0	2.23418
		9.0	1.82664				9.0	2.09173
		10.0	1.73639				10.0	1.97158
		15.0	1.43650				15.0	1.57679
	0.140	2.0	4.02480	0.800	0.010	0.080	2.0	2.78340
		3.0	3.23621				3.0	2.33475
		4.0	2.79469				4.0	2.08647
		5.0	2.49912				5.0	1.92092
		6.0	2.28140				6.0	1.79864
		7.0	2.11161				7.0	1.70251
		8.0	1.97427				8.0	1.62384
		9.0	1.86037				9.0	1.55768
		10.0	1.76423				10.0	1.50096
		15.0	1.44711				15.0	1.30539
	0.160	2.0	4.14741			0.100	2.0	2.92424
		3.0	3.33572				3.0	2.45669
		4.0	2.87542				4.0	2.19199
		5.0	2.56458				5.0	2.01218
		6.0	2.33446				6.0	1.87751
		7.0	2.15460				7.0	1.77064
		8.0	2.00908				8.0	1.68266
		9.0	1.88855				9.0	1.60843
		10.0	1.78703				10.0	1.54473
		15.0	1.45496				15.0	1.32610

TABLE OF FACTORS

$N = 0.0$

V	C4 K 0.800	LIFE	VALUE	V	C4 K 0.800	LIFE	VALUE
0.010	0.120	2.0	3.05955	0.030	0.080	2.0	3.56556
		3.0	2.57153			3.0	2.96294
		4.0	2.28939			4.0	2.62381
		5.0	2.09474			5.0	2.39428
		6.0	1.94746			6.0	2.22260
		7.0	1.82987			7.0	2.08625
		8.0	1.73279			8.0	1.97372
		9.0	1.65083			9.0	1.87842
		10.0	1.58057			10.0	1.79622
		15.0	1.34144			15.0	1.50878
	0.140	2.0	3.18956		0.100	2.0	3.74598
		3.0	2.67968			3.0	3.11768
		4.0	2.37931			4.0	2.75650
		5.0	2.16946			5.0	2.50802
		6.0	2.00951			6.0	2.32007
		7.0	1.88137			7.0	2.16974
		8.0	1.77550			8.0	2.04522
		9.0	1.68624			9.0	1.93963
		10.0	1.60991			10.0	1.84860
		15.0	1.35281			15.0	1.53271
	0.160	2.0	3.31447		0.120	2.0	3.91932
		3.0	2.78153			3.0	3.26342
		4.0	2.46231			4.0	2.87899
		5.0	2.23706			5.0	2.61094
		6.0	2.06453			6.0	2.40651
		7.0	1.92613			7.0	2.24233
		8.0	1.81190			8.0	2.10615
		9.0	1.71582			9.0	1.99075
		10.0	1.63393			10.0	1.89149
		15.0	1.36123			15.0	1.55045
0.020	0.080	2.0	3.12246		0.140	2.0	4.08586
		3.0	2.60696			3.0	3.40066
		4.0	2.31920			4.0	2.99205
		5.0	2.12581			5.0	2.70406
		6.0	1.98200			6.0	2.48318
		7.0	1.86833			7.0	2.30543
		8.0	1.77487			8.0	2.15807
		9.0	1.69596			9.0	2.03346
		10.0	1.62808			10.0	1.92660
		15.0	1.39220			15.0	1.56358
	0.100	2.0	3.28045		0.160	2.0	4.24587
		3.0	2.74312			3.0	3.52992
		4.0	2.43649			4.0	3.09643
		5.0	2.22680			5.0	2.78832
		6.0	2.06892			6.0	2.55117
		7.0	1.94310			7.0	2.36029
		8.0	1.83916			8.0	2.20231
		9.0	1.75122			9.0	2.06913
		10.0	1.67556			10.0	1.95535
		15.0	1.41429			15.0	1.57332
	0.120	2.0	3.43225				
		3.0	2.87134				
		4.0	2.54475				
		5.0	2.31817	FACTOR			VALUE
		6.0	2.14600				
		7.0	2.00810	S10			0.00000
		8.0	1.89395				
		9.0	1.79738				
		10.0	1.71444				
		15.0	1.43065				
	0.140	2.0	3.57809	FACTOR	T3	LIFE	VALUE
		3.0	2.99210	F11	2.000	2.0	0.00000
		4.0	2.64470			3.0	3.05086
		5.0	2.40085			4.0	2.40409
		6.0	2.21437			5.0	3.68736
		7.0	2.06461			6.0	3.22260
		8.0	1.94064			7.0	3.97660
		9.0	1.83594			8.0	3.64258
		10.0	1.74627			9.0	4.14831
		15.0	1.44278			10.0	3.90150
						15.0	4.41833
	0.160	2.0	3.71822				
		3.0	3.10583				
		4.0	2.73696				
		5.0	2.47566				
		6.0	2.27501				
		7.0	2.11373				
		8.0	1.98042				
		9.0	1.86814				
		10.0	1.77232				
		15.0	1.45176				

CONSTRUCTION EQUIPMENT POLICY

N = 0.0

FACTOR	T3	LIFE	VALUE
F11	3.000	2.0	0.00000
		3.0	0.00000
		4.0	2.18553
		5.0	1.83534
		6.0	1.60401
		7.0	2.52292
		8.0	2.31100
		9.0	2.14857
		10.0	2.67204
		15.0	2.59275
	4.000	2.0	0.00000
		3.0	0.00000
		4.0	0.00000
		5.0	1.66849
		6.0	1.45819
		7.0	1.30962
		8.0	1.19962
		9.0	1.87707
		10.0	1.76538
		15.0	1.84917

FACTOR	VALUE
D1	0.00000

FACTOR	V	LIFE	VALUE
D2	0.010	3.0	3.90292
		4.0	2.84867
		5.0	2.24683
		6.0	1.85802
		7.0	1.58629
		8.0	1.38577
		9.0	1.23178
		10.0	1.10986
		15.0	0.75124
	0.020	3.0	4.35797
		4.0	3.16642
		5.0	2.48648
		6.0	2.04744
		7.0	1.74079
		8.0	1.51466
		9.0	1.34113
		10.0	1.20386
		15.0	0.80120
	0.030	3.0	4.95304
		4.0	3.58230
		5.0	2.80050
		6.0	2.29598
		7.0	1.94384
		8.0	1.68436
		9.0	1.48542
		10.0	1.32818
		15.0	0.86828

FACTOR	VALUE
D5	0.00000

FACTOR	K	V	C4	LIFE	VALUE
D6	0.700	0.010	0.080	2.0	2.15871
				3.0	1.55485
				4.0	1.24513
				5.0	1.05366
				6.0	0.92175
				7.0	0.82422
				8.0	0.74844
				9.0	0.68740
				10.0	0.63683
				15.0	0.47136
			0.100	2.0	2.29731
				3.0	1.68232
				4.0	1.36244
				5.0	1.16166
				6.0	1.02124
				7.0	0.91592
				8.0	0.83300
				9.0	0.76541
				10.0	0.70885
				15.0	0.51996
			0.120	2.0	2.43047
				3.0	1.80237
				4.0	1.47072
				5.0	1.25939
				6.0	1.10948
				7.0	0.99564
				8.0	0.90506
				9.0	0.83058
				10.0	0.76781
				15.0	0.55596
			0.140	2.0	2.55842
				3.0	1.91543
				4.0	1.57069
				5.0	1.34782
				6.0	1.18775
				7.0	1.06494
				8.0	0.96646
				9.0	0.88500
				10.0	0.81608
				15.0	0.58264
			0.160	2.0	2.68134
				3.0	2.02191
				4.0	1.66296
				5.0	1.42783
				6.0	1.25716
				7.0	1.12519
				8.0	1.01879
				9.0	0.93047
				10.0	0.85560
				15.0	0.60240
		0.020	0.080	2.0	2.42167
				3.0	1.73613
				4.0	1.38402
				5.0	1.16604
				6.0	1.01572
				7.0	0.90449
				8.0	0.81805
				9.0	0.74842
				10.0	0.69077
				15.0	0.50271
			0.100	2.0	2.57715
				3.0	1.87847
				4.0	1.51441
				5.0	1.28557
				6.0	1.12535
				7.0	1.00512
				8.0	0.91048
				9.0	0.83336
				10.0	0.76889
				15.0	0.55454
			0.120	2.0	2.72654
				3.0	2.01251
				4.0	1.63477
				5.0	1.39372
				6.0	1.22259
				7.0	1.09261
				8.0	0.98924
				9.0	0.90431
				10.0	0.83284
				15.0	0.59294

TABLE OF FACTORS

N = 0.0

V	C4	LIFE	VALUE	K	V	C4	LIFE	VALUE
0.020	0.140	2.0	2.87007	0.700	0.030	0.160	2.0	3.43483
		3.0	2.13876				3.0	2.56592
		4.0	1.74588				4.0	2.09123
		5.0	1.49158				5.0	1.77968
		6.0	1.30884				6.0	1.55350
		7.0	1.16866				7.0	1.37881
		8.0	1.05635				8.0	1.23831
		9.0	0.96357				9.0	1.12206
		10.0	0.88520				10.0	1.02391
		15.0	0.62138				15.0	0.69625
	0.160	2.0	3.00797	0.800	0.010	0.080	2.0	1.65374
		3.0	2.25765				3.0	1.24214
		4.0	1.84845				4.0	1.02716
		5.0	1.58012				5.0	0.89152
		6.0	1.38533				6.0	0.79605
		7.0	1.23478				7.0	0.72394
		8.0	1.11354				8.0	0.66674
		9.0	1.01307				9.0	0.61974
		10.0	0.92807				10.0	0.58008
		15.0	0.64246				15.0	0.44457
0.030	0.080	2.0	2.76533			0.100	2.0	1.81214
		3.0	1.97319				3.0	1.38782
		4.0	1.56579				4.0	1.16123
		5.0	1.31330				5.0	1.01496
		6.0	1.13902				6.0	0.90976
		7.0	1.00999				7.0	0.82874
		8.0	0.90971				8.0	0.76338
		9.0	0.82894				9.0	0.70890
		10.0	0.76211				10.0	0.66238
		15.0	0.54480				15.0	0.50012
	0.100	2.0	2.94287			0.120	2.0	1.96433
		3.0	2.13496				3.0	1.52502
		4.0	1.71331				4.0	1.28499
		5.0	1.44792				5.0	1.12664
		6.0	1.26196				6.0	1.01061
		7.0	1.12236				7.0	0.91985
		8.0	1.01249				8.0	0.84573
		9.0	0.92302				9.0	0.78337
		10.0	0.84829				10.0	0.72976
		15.0	0.60097				15.0	0.54126
	0.120	2.0	3.11346			0.140	2.0	2.11055
		3.0	2.28732				3.0	1.65424
		4.0	1.84948				4.0	1.39923
		5.0	1.56973				5.0	1.22770
		6.0	1.37101				6.0	1.10005
		7.0	1.22005				7.0	0.99906
		8.0	1.10007				8.0	0.91591
		9.0	1.00160				9.0	0.84558
		10.0	0.91885				10.0	0.78493
		15.0	0.64258				15.0	0.57175
	0.140	2.0	3.27736			0.160	2.0	2.25104
		3.0	2.43080				3.0	1.77592
		4.0	1.97519				4.0	1.50469
		5.0	1.67995				5.0	1.31915
		6.0	1.46772				6.0	1.17939
		7.0	1.30498				7.0	1.06791
		8.0	1.17471				8.0	0.97571
		9.0	1.06724				9.0	0.89754
		10.0	0.97661				10.0	0.83010
		15.0	0.67341				15.0	0.59433
					0.020	0.080	2.0	1.85519
							3.0	1.38696
							4.0	1.14174
							5.0	0.98661
							6.0	0.87721
							7.0	0.79445
							8.0	0.72875
							9.0	0.67476
							10.0	0.62921
							15.0	0.47414

$N = 0.0$

V	FACTOR D6 C4	K 0.800 LIFE	VALUE	V	C4	LIFE	VALUE
0.020	0.100	2.0	2.03289	0.030	0.140	2.0	2.70364
		3.0	1.54963			3.0	2.09932
		4.0	1.29075			4.0	1.75957
		5.0	1.12321			5.0	1.53024
		6.0	1.00251			6.0	1.35935
		7.0	0.90946			7.0	1.22425
		8.0	0.83438			8.0	1.11326
		9.0	0.77183			9.0	1.01969
		10.0	0.71848			10.0	0.93934
		15.0	0.53337			15.0	0.66083
	0.120	2.0	2.20362		0.160	2.0	2.88361
		3.0	1.70283			3.0	2.25375
		4.0	1.42832			4.0	1.89219
		5.0	1.24681			5.0	1.64421
		6.0	1.11364			6.0	1.45739
		7.0	1.00944			7.0	1.30862
		8.0	0.92439			8.0	1.18594
		9.0	0.85292			9.0	1.08235
		10.0	0.79157			10.0	0.99339
		15.0	0.57726			15.0	0.68693
	0.140	2.0	2.36765				
		3.0	1.84711				
		4.0	1.55530				
		5.0	1.35865	FACTOR			VALUE
		6.0	1.21220				
		7.0	1.09636	T1			0.00000
		8.0	1.00109				
		9.0	0.92064				
		10.0	0.85141				
		15.0	0.60977				
	0.160	2.0	2.52525	FACTOR	V	LIFE	VALUE
		3.0	1.98298				
		4.0	1.67252	T2	0.010	2.0	1.13869
		5.0	1.45985			3.0	1.12315
		6.0	1.29962			4.0	1.10836
		7.0	1.17193			5.0	0.92281
		8.0	1.06646			6.0	0.79986
		9.0	0.97721			7.0	0.71267
		10.0	0.90041			8.0	0.64783
		15.0	0.63385			9.0	0.59788
						10.0	0.55835
						15.0	0.44390
0.030	0.080	2.0	2.11846				
		3.0	1.57634		0.020	2.0	1.27740
		4.0	1.29169			3.0	1.25409
		5.0	1.11121			4.0	1.23199
		6.0	0.98370			5.0	1.02124
		7.0	0.88712			6.0	0.88140
		8.0	0.81040			7.0	0.78208
		9.0	0.74735			8.0	0.70808
		10.0	0.69419			9.0	0.65095
		15.0	0.51384			10.0	0.60564
						15.0	0.47343
	0.100	2.0	2.32137				
		3.0	1.76123		0.030	2.0	1.45868
		4.0	1.46028			3.0	1.42534
		5.0	1.26506			4.0	1.39380
		6.0	1.12421			5.0	1.15021
		7.0	1.01554			6.0	0.98840
		8.0	0.92787			7.0	0.87331
		9.0	0.85487			8.0	0.78741
		10.0	0.79268			9.0	0.72099
		15.0	0.57803			10.0	0.66818
						15.0	0.51307
	0.120	2.0	2.51633				
		3.0	1.93535				
		4.0	1.61591				
		5.0	1.40427				
		6.0	1.24883	FACTOR			VALUE
		7.0	1.12718				
		8.0	1.02796	G1			0.00000
		9.0	0.94468				
		10.0	0.87332				
		15.0	0.62559				

TABLE OF FACTORS 257

$N = 0.0$

FACTOR	K	V	C4	LIFE	VALUE	V	C4	LIFE	VALUE
G2	0.700	0.010	0.080	3.0	1.71608	0.020	0.120	3.0	1.68483
				4.0	1.06455			4.0	0.98593
				5.0	0.71323			5.0	0.62099
				6.0	0.50060			6.0	0.40813
				7.0	0.36222			7.0	0.27519
				8.0	0.26765			8.0	0.18835
				9.0	0.20077			9.0	0.12985
				10.0	0.15224			10.0	0.08957
				15.0	0.04120			15.0	0.01030
			0.100	3.0	1.60939		0.140	3.0	1.57917
				4.0	0.97222			4.0	0.89848
				5.0	0.63338			5.0	0.54864
				6.0	0.43158			6.0	0.34831
				7.0	0.30260			7.0	0.22575
				8.0	0.21618			8.0	0.14750
				9.0	0.15636			9.0	0.09611
				10.0	0.11394			10.0	0.06172
				15.0	0.02308			15.0	-0.00031
			0.120	3.0	1.50890		0.160	3.0	1.47966
				4.0	0.88699			4.0	0.81775
				5.0	0.56114			5.0	0.48318
				6.0	0.37037			6.0	0.29525
				7.0	0.25077			7.0	0.18276
				8.0	0.17232			8.0	0.11269
				9.0	0.11926			9.0	0.06793
				10.0	0.08258			10.0	0.03892
				15.0	0.00965			15.0	-0.00817
			0.140	3.0	1.41427	0.030	0.080	3.0	2.17781
				4.0	0.80832			4.0	1.33870
				5.0	0.49576			5.0	0.88898
				6.0	0.31609			6.0	0.61859
				7.0	0.20571			7.0	0.44386
				8.0	0.13495			8.0	0.32533
				9.0	0.08827			9.0	0.24211
				10.0	0.05691			10.0	0.18219
				15.0	-0.00029			15.0	0.04762
			0.160	3.0	1.32515		0.100	3.0	2.04241
				4.0	0.73569			4.0	1.22260
				5.0	0.43661			5.0	0.78946
				6.0	0.26794			6.0	0.53331
				7.0	0.16654			7.0	0.37080
				8.0	0.10310			8.0	0.26276
				9.0	0.06239			9.0	0.18855
				10.0	0.03588			10.0	0.13635
				15.0	-0.00766			15.0	0.02668
		0.020	0.080	3.0	1.91616		0.120	3.0	1.91489
				4.0	1.18329			4.0	1.11542
				5.0	0.78930			5.0	0.69941
				6.0	0.55163			6.0	0.45768
				7.0	0.39749			7.0	0.30729
				8.0	0.29255			8.0	0.20945
				9.0	0.21859			9.0	0.14382
				10.0	0.16513			10.0	0.09882
				15.0	0.04394			15.0	0.01116
			0.100	3.0	1.79703		0.140	3.0	1.79480
				4.0	1.08066			4.0	1.01649
				5.0	0.70094			5.0	0.61793
				6.0	0.47558			6.0	0.39059
				7.0	0.33207			7.0	0.25208
				8.0	0.23629			8.0	0.16403
				9.0	0.17024			9.0	0.10645
				10.0	0.12359			10.0	0.06810
				15.0	0.02461			15.0	-0.00034
							0.160	3.0	1.68170
								4.0	0.92516
								5.0	0.54420
								6.0	0.33109
								7.0	0.20408
								8.0	0.12531
								9.0	0.07524
								10.0	0.04294
								15.0	-0.00885

$N = 0.0$

K	FACTOR V	G2 C4	LIFE	VALUE	V	C4	LIFE	VALUE
0.800	0.010	0.080	3.0	1.97782		0.140	3.0	1.82327
			4.0	1.23610			4.0	1.04848
			5.0	0.83309			5.0	0.64691
			6.0	0.58778			6.0	0.41534
			7.0	0.42741			7.0	0.27275
			8.0	0.31739			8.0	0.18113
			9.0	0.23928			9.0	0.12055
			10.0	0.18243			10.0	0.07969
			15.0	0.05119			15.0	0.00402
		0.100	3.0	1.85588		0.160	3.0	1.70955
			4.0	1.13058			4.0	0.95622
			5.0	0.74184			5.0	0.57210
			6.0	0.50891			6.0	0.35470
			7.0	0.35927			7.0	0.22363
			8.0	0.25856			8.0	0.14135
			9.0	0.18853			9.0	0.08834
			10.0	0.13865			10.0	0.05364
			15.0	0.03048			15.0	-0.00496
		0.120	3.0	1.74104	0.030	0.080	3.0	2.50996
			4.0	1.03318			4.0	1.55443
			5.0	0.65927			5.0	1.03838
			6.0	0.43896			6.0	0.72633
			7.0	0.30004			7.0	0.52374
			8.0	0.20844			8.0	0.38577
			9.0	0.14613			9.0	0.28856
			10.0	0.10281			10.0	0.21831
			15.0	0.01514			15.0	0.05917
		0.140	3.0	1.63289		0.100	3.0	2.35522
			4.0	0.94326			4.0	1.42174
			5.0	0.58456			5.0	0.92464
			6.0	0.37691			6.0	0.62887
			7.0	0.24855			7.0	0.44025
			8.0	0.16572			8.0	0.31427
			9.0	0.11072			9.0	0.22735
			10.0	0.07347			10.0	0.16593
			15.0	0.00377			15.0	0.03523
		0.160	3.0	1.53104		0.120	3.0	2.20948
			4.0	0.86026			4.0	1.29925
			5.0	0.51696			5.0	0.82173
			6.0	0.32189			6.0	0.54243
			7.0	0.20378			7.0	0.36767
			8.0	0.12932			8.0	0.25335
			9.0	0.08114			9.0	0.17622
			10.0	0.04945			10.0	0.12304
			15.0	-0.00465			15.0	0.01750
	0.020	0.080	3.0	2.20841		0.140	3.0	2.07224
			4.0	1.37397			4.0	1.18618
			5.0	0.92195			5.0	0.72861
			6.0	0.64771			6.0	0.46576
			7.0	0.46903			7.0	0.30457
			8.0	0.34690			8.0	0.20143
			9.0	0.26053			9.0	0.13352
			10.0	0.19788			10.0	0.08792
			15.0	0.05459			15.0	0.00436
		0.100	3.0	2.07226		0.160	3.0	1.94298
			4.0	1.25669			4.0	1.08181
			5.0	0.82096			5.0	0.64435
			6.0	0.56079			6.0	0.39776
			7.0	0.39426			7.0	0.24971
			8.0	0.28261			8.0	0.15719
			9.0	0.20526			9.0	0.09785
			10.0	0.15040			10.0	0.05917
			15.0	0.03251			15.0	-0.00537
		0.120	3.0	1.94403				
			4.0	1.14842				
			5.0	0.72959				
			6.0	0.48371				
			7.0	0.32926				
			8.0	0.22782				
			9.0	0.15910				
			10.0	0.11152				
			15.0	0.01614				

FACTOR	VALUE
TC11	1.00000

FACTOR	VALUE
TC12	1.00000

TABLE OF FACTORS

FACTOR	K	V	N = 0.0 C4	LIFE	VALUE
IC2	0.700	0.010	0.080		
				3.0	1.59649
				4.0	1.25803
				5.0	1.04079
				6.0	0.88582
				7.0	0.76770
				8.0	0.67357
				9.0	0.59619
				10.0	0.53114
				15.0	0.31538
			0.100		
				3.0	1.70319
				4.0	1.35035
				5.0	1.12063
				6.0	0.95484
				7.0	0.82732
				8.0	0.72504
				9.0	0.64060
				10.0	0.56944
				15.0	0.33350
			0.120		
				3.0	1.80367
				4.0	1.43558
				5.0	1.19288
				6.0	1.01605
				7.0	0.87914
				8.0	0.76890
				9.0	0.67770
				10.0	0.60080
				15.0	0.34693
			0.140		
				3.0	1.89830
				4.0	1.51425
				5.0	1.25825
				6.0	1.07033
				7.0	0.92420
				8.0	0.80628
				9.0	0.70869
				10.0	0.62648
				15.0	0.35687
			0.160		
				3.0	1.98742
				4.0	1.58688
				5.0	1.31740
				6.0	1.11849
				7.0	0.96337
				8.0	0.83813
				9.0	0.73457
				10.0	0.64750
				15.0	0.36424
		0.020	0.080		
				3.0	1.89921
				4.0	1.50989
				5.0	1.25846
				6.0	1.07808
				7.0	0.93987
				8.0	0.82922
				9.0	0.73789
				10.0	0.66083
				15.0	0.40286

CONSTRUCTION EQUIPMENT POLICY

$N = 0.0$

V	FACTOR C4	IC2	K 0.080 LIFE	VALUE	K	V	C4	LIFE	VALUE
0.020	0.100				0.700	0.030	0.140		
			3.0	2.01835				3.0	2.67811
			4.0	1.61251				4.0	2.16176
			5.0	1.34682				5.0	1.81474
			6.0	1.15413				6.0	1.55834
			7.0	1.00529				7.0	1.35792
			8.0	0.88548				8.0	1.19547
			9.0	0.78625				9.0	1.06052
			10.0	0.70237				10.0	0.94643
			15.0	0.42218				15.0	0.56828
	0.120						0.160		
			3.0	2.13055				3.0	2.79121
			4.0	1.70725				4.0	2.25309
			5.0	1.42678				5.0	1.88846
			6.0	1.22158				6.0	1.61784
			7.0	1.06217				7.0	1.40592
			8.0	0.93342				8.0	1.23419
			9.0	0.82664				9.0	1.09173
			10.0	0.73639				10.0	0.97158
			15.0	0.43650				15.0	0.57679
	0.140				0.800	0.010	0.080		
			3.0	2.23621				3.0	1.33475
			4.0	1.79470				4.0	1.08648
			5.0	1.49912				5.0	0.92093
			6.0	1.28140				6.0	0.79864
			7.0	1.11161				7.0	0.70251
			8.0	0.97427				8.0	0.62384
			9.0	0.86038				9.0	0.55768
			10.0	0.76423				10.0	0.50096
			15.0	0.44711				15.0	0.30539
	0.160						0.100		
			3.0	2.33572				3.0	1.45669
			4.0	1.87542				4.0	1.19199
			5.0	1.56458				5.0	1.01218
			6.0	1.33446				6.0	0.87751
			7.0	1.15460				7.0	0.77064
			8.0	1.00908				8.0	0.68266
			9.0	0.88856				9.0	0.60843
			10.0	0.78704				10.0	0.54473
			15.0	0.45497				15.0	0.32610
0.030	0.080						0.120		
			3.0	2.29510				3.0	1.57153
			4.0	1.83954				4.0	1.28939
			5.0	1.54368				5.0	1.09474
			6.0	1.33034				6.0	0.94746
			7.0	1.16614				7.0	0.82987
			8.0	1.03417				8.0	0.73279
			9.0	0.92486				9.0	0.65083
			10.0	0.83234				10.0	0.58057
			15.0	0.52032				15.0	0.34145
	0.100						0.140		
			3.0	2.43050				3.0	1.67968
			4.0	1.95565				4.0	1.37931
			5.0	1.64320				5.0	1.16946
			6.0	1.41562				6.0	1.00951
			7.0	1.23919				7.0	0.88137
			8.0	1.09673				8.0	0.77550
			9.0	0.97842				9.0	0.68624
			10.0	0.87818				10.0	0.60991
			15.0	0.54127				15.0	0.35281
	0.120						0.160		
			3.0	2.55802				3.0	1.78153
			4.0	2.06282				4.0	1.46231
			5.0	1.73325				5.0	1.23706
			6.0	1.49126				6.0	1.06453
			7.0	1.30270				7.0	0.92614
			8.0	1.15004				8.0	0.81190
			9.0	1.02316				9.0	0.71582
			10.0	0.91570				10.0	0.63394
			15.0	0.55678				15.0	0.36123

TABLE OF FACTORS

$N = 0.0$

V	C4	LIFE	VALUE		C4	LIFE	VALUE
0.020	0.080				0.120		
		3.0	1.60696			3.0	2.26342
		4.0	1.31920			4.0	1.87899
		5.0	1.12581			5.0	1.61094
		6.0	0.98200			6.0	1.40651
		7.0	0.86833			7.0	1.24233
		8.0	0.77487			8.0	1.10615
		9.0	0.69596			9.0	0.99075
		10.0	0.62808			10.0	0.89149
		15.0	0.39220			15.0	0.55045
	0.100				0.140		
		3.0	1.74312			3.0	2.40067
		4.0	1.43649			4.0	1.99206
		5.0	1.22680			5.0	1.70406
		6.0	1.06892			6.0	1.48318
		7.0	0.94310			7.0	1.30543
		8.0	0.83916			8.0	1.15807
		9.0	0.75122			9.0	1.03346
		10.0	0.67556			10.0	0.92660
		15.0	0.41429			15.0	0.56358
	0.120				0.160		
		3.0	1.87134			3.0	2.52992
		4.0	1.54476			4.0	2.09644
		5.0	1.31817			5.0	1.78832
		6.0	1.14600			6.0	1.55117
		7.0	1.00810			7.0	1.36029
		8.0	0.89395			8.0	1.20231
		9.0	0.79738			9.0	1.06913
		10.0	0.71444			10.0	0.95535
		15.0	0.43065			15.0	0.57332
	0.140						
		3.0	1.99210				
		4.0	1.64470				
		5.0	1.40085				
		6.0	1.21437				
		7.0	1.06461				
		8.0	0.94064				
		9.0	0.83594				
		10.0	0.74627				
		15.0	0.44278				
	0.160						
		3.0	2.10583				
		4.0	1.73696				
		5.0	1.47566				
		6.0	1.27501				
		7.0	1.11373				
		8.0	0.98042				
		9.0	0.86814				
		10.0	0.77232				
		15.0	0.45176				
0.030	0.080						
		3.0	1.96294				
		4.0	1.62381				
		5.0	1.39428				
		6.0	1.22260				
		7.0	1.08625				
		8.0	0.97373				
		9.0	0.87842				
		10.0	0.79622				
		15.0	0.50878				
	0.100						
		3.0	2.11769				
		4.0	1.75650				
		5.0	1.50802				
		6.0	1.32007				
		7.0	1.16975				
		8.0	1.04522				
		9.0	0.93963				
		10.0	0.84860				
		15.0	0.53271				

ALL FACTORS FOR N=1.0 ARE COMPUTED TO HAVE THE FOLLOWING VALUES AT 10% RATE OF RETURN

$N = 1.0$

FACTOR	S1		VALUE
F1	0.020		0.94450
	0.040		0.93530
	0.060		0.92621
	0.080		0.91725
	0.100		0.90840

FACTOR	S2	LIFE	VALUE
F2	0.020	2.0	9.35595
		3.0	9.27096
		4.0	9.18995
		5.0	9.11278
		6.0	9.03935
		7.0	8.96953
		8.0	8.90321
		9.0	8.84027
		10.0	8.78059
		15.0	8.52701

N = 1.0

FACTOR	S2	LIFE	VALUE
R2	0.040	2.0	9.17834
		3.0	9.01385
		4.0	8.85903
		5.0	8.71343
		6.0	8.57658
		7.0	8.44807
		8.0	8.32746
		9.0	8.21436
		10.0	8.10838
		15.0	7.67266
	0.060	2.0	9.00531
		3.0	8.76648
		4.0	8.54453
		5.0	8.33836
		6.0	8.14696
		7.0	7.96936
		8.0	7.80464
		9.0	7.65195
		10.0	7.51050
		15.0	6.94667
	0.080	2.0	8.83670
		3.0	8.52843
		4.0	8.24550
		5.0	7.98590
		6.0	7.74776
		7.0	7.52937
		8.0	7.32912
		9.0	7.14557
		10.0	6.97737
		15.0	6.32615
	0.100	2.0	8.67240
		3.0	8.29929
		4.0	7.96107
		5.0	7.65447
		6.0	7.37650
		7.0	7.12447
		8.0	6.89594
		9.0	6.68870
		10.0	6.50075
		15.0	5.79271

FACTOR	G	S2	LIFE	VALUE
R3	0.020	0.020	2.0	7.72778
			3.0	7.72778
			4.0	7.72778
			5.0	7.72778
			6.0	7.72778
			7.0	7.72778
			8.0	7.72778
			9.0	7.72778
			10.0	7.72778
			15.0	7.72778
		0.040	2.0	7.58108
			3.0	7.51346
			4.0	7.44951
			5.0	7.38912
			6.0	7.33216
			7.0	7.27850
			8.0	7.22804
			9.0	7.18063
			10.0	7.13617
			15.0	6.95351
		0.060	2.0	7.43816
			3.0	7.30727
			4.0	7.18505
			5.0	7.07106
			6.0	6.96487
			7.0	6.86607
			8.0	6.77424
			9.0	6.68900
			10.0	6.60997
			15.0	6.29556
		0.080	2.0	7.29890
			3.0	7.10884
			4.0	6.93360
			5.0	6.77216
			6.0	6.62359
			7.0	6.48699
			8.0	6.36151
			9.0	6.24635
			10.0	6.14076
			15.0	5.73320

G	S2	LIFE	VALUE
	0.100	2.0	7.16319
		3.0	6.91784
		4.0	6.69442
		5.0	6.49111
		6.0	6.30620
		7.0	6.13815
		8.0	5.98551
		9.0	5.84697
		10.0	5.72130
		15.0	5.24976
0.040	0.020	2.0	6.58005
		3.0	6.63927
		4.0	6.69626
		5.0	6.75099
		6.0	6.80344
		7.0	6.85359
		8.0	6.90144
		9.0	6.94701
		10.0	6.99029
		15.0	7.17392
	0.040	2.0	6.45514
		3.0	6.45514
		4.0	6.45514
		5.0	6.45514
		6.0	6.45514
		7.0	6.45514
		8.0	6.45514
		9.0	6.45514
		10.0	6.45514
		15.0	6.45514
	0.060	2.0	6.33345
		3.0	6.27799
		4.0	6.22598
		5.0	6.17728
		6.0	6.13179
		7.0	6.08936
		8.0	6.04987
		9.0	6.01319
		10.0	5.97916
		15.0	5.84435
	0.080	2.0	6.21487
		3.0	6.10751
		4.0	6.00809
		5.0	5.91617
		6.0	5.83133
		7.0	5.75317
		8.0	5.68127
		9.0	5.61525
		10.0	5.55473
		15.0	5.32230
	0.100	2.0	6.09932
		3.0	5.94342
		4.0	5.80084
		5.0	5.67064
		6.0	5.55190
		7.0	5.44379
		8.0	5.34548
		9.0	5.25622
		10.0	5.17529
		15.0	4.87350
0.060	0.020	2.0	5.72717
		3.0	5.82975
		4.0	5.92891
		5.0	6.02449
		6.0	6.11634
		7.0	6.20436
		8.0	6.28846
		9.0	6.36859
		10.0	6.44474
		15.0	6.76660

TABLE OF FACTORS   263

$N = 1.0$

G	S2	LIFE	VALUE		S2	LIFE	VALUE
0.060	0.040	2.0	5.61844		0.080	2.0	4.78693
		3.0	5.66807			3.0	4.78693
		4.0	5.71542			4.0	4.78693
		5.0	5.76048			5.0	4.78693
		6.0	5.80322			6.0	4.78693
		7.0	5.84366			7.0	4.78693
		8.0	5.88179			8.0	4.78693
		9.0	5.91768			9.0	4.78693
		10.0	5.95136			10.0	4.78693
		15.0	6.08864			15.0	4.78693
	0.060	2.0	5.51252		0.100	2.0	4.69793
		3.0	5.51252			3.0	4.65832
		4.0	5.51252			4.0	4.62181
		5.0	5.51252			5.0	4.58826
		6.0	5.51252			6.0	4.55755
		7.0	5.51252			7.0	4.52951
		8.0	5.51252			8.0	4.50400
		9.0	5.51252			9.0	4.48086
		10.0	5.51252			10.0	4.45994
		15.0	5.51252			15.0	4.38328
	0.080	2.0	5.40932				
		3.0	5.36283				
		4.0	5.31960				
		5.0	5.27951	FACTOR			VALUE
		6.0	5.24241				
		7.0	5.20818	F1			0.95382
		8.0	5.17666				
		9.0	5.14772				
		10.0	5.12122				
		15.0	5.02011	FACTOR	W1		VALUE
	0.100	2.0	5.30874				
		3.0	5.21875	F2	0.020		0.94450
		4.0	5.13610				
		5.0	5.06040		0.040		0.93530
		6.0	4.99120				
		7.0	4.92810		0.060		0.92621
		8.0	4.87069				
		9.0	4.81858		0.080		0.91725
		10.0	4.77139				
		15.0	4.59680		0.100		0.90840
0.080	0.020	2.0	5.06821		0.120		0.89967
		3.0	5.20371				
		4.0	5.33523				
		5.0	5.46241				
		6.0	5.58493				
		7.0	5.70254	FACTOR		LIFE	VALUE
		8.0	5.81503				
		9.0	5.92223	F3		2.0	8.62964
		10.0	6.02406			3.0	8.66928
		15.0	6.45230			4.0	8.70766
						5.0	8.74477
	0.040	2.0	4.97200			6.0	8.78061
		3.0	5.05939			7.0	8.81517
		4.0	5.14312			8.0	8.84846
		5.0	5.22303			9.0	8.88049
		6.0	5.29902			10.0	8.91126
		7.0	5.37101			15.0	9.04673
		8.0	5.43898				
		9.0	5.50292				
		10.0	5.56288				
		15.0	5.80582	FACTOR		LIFE	VALUE
	0.060	2.0	4.87826				
		3.0	4.92055	F4		2.0	8.02497
		4.0	4.96053			3.0	8.09927
		5.0	4.99821			4.0	8.17139
		6.0	5.03358			5.0	8.24128
		7.0	5.06666			6.0	8.30890
		8.0	5.09750			7.0	8.37421
		9.0	5.12616			8.0	8.43721
		10.0	5.15269			9.0	8.49787
		15.0	5.25647			10.0	8.55619
						15.0	8.81309

N = 1.0

FACTOR	W2	LIFE	VALUE
F5	0.020	2.0	7.87159
		3.0	7.87231
		4.0	7.87300
		5.0	7.87366
		6.0	7.87429
		7.0	7.87490
		8.0	7.87547
		9.0	7.87601
		10.0	7.87653
		15.0	7.87873
	0.040	2.0	7.72216
		3.0	7.65398
		4.0	7.58951
		5.0	7.52860
		6.0	7.47117
		7.0	7.41707
		8.0	7.36618
		9.0	7.31837
		10.0	7.27353
		15.0	7.08933
	0.060	2.0	7.57657
		3.0	7.44393
		4.0	7.32007
		5.0	7.20454
		6.0	7.09692
		7.0	6.99677
		8.0	6.90371
		9.0	6.81731
		10.0	6.73721
		15.0	6.41853
	0.080	2.0	7.43472
		3.0	7.24179
		4.0	7.06389
		5.0	6.90001
		6.0	6.74917
		7.0	6.61048
		8.0	6.48308
		9.0	6.36617
		10.0	6.25897
		15.0	5.84519
	0.100	2.0	7.29649
		3.0	7.04722
		4.0	6.82023
		5.0	6.61364
		6.0	6.42576
		7.0	6.25500
		8.0	6.09990
		9.0	5.95912
		10.0	5.83143
		15.0	5.35231
	0.120	2.0	7.16178
		3.0	6.85991
		4.0	6.58837
		5.0	6.34421
		6.0	6.12471
		7.0	5.92747
		8.0	5.75026
		9.0	5.59111
		10.0	5.44823
		15.0	4.92614

FACTOR	C3	LIFE	VALUE
F6	0.080		0.16080
	0.100		0.17742
	0.120		0.19371
	0.140		0.20967
	0.160		0.22532

FACTOR	K	V	C4	LIFE	VALUE
F7	0.700	0.010	0.080	2.0	2.96806
				3.0	2.38416
				4.0	2.07338
				5.0	1.87391
				6.0	1.73161
				7.0	1.62315
				8.0	1.53672
				9.0	1.46567
				10.0	1.40594
				15.0	1.20782
			0.100	2.0	3.08121
				3.0	2.48213
				4.0	2.15816
				5.0	1.94722
				6.0	1.79498
				7.0	1.67789
				8.0	1.58398
				9.0	1.50645
				10.0	1.44111
				15.0	1.22446
			0.120	2.0	3.18993
				3.0	2.57440
				4.0	2.23641
				5.0	2.01356
				6.0	1.85119
				7.0	1.72548
				8.0	1.62425
				9.0	1.54051
				10.0	1.46990
				15.0	1.23679
			0.140	2.0	3.29438
				3.0	2.66129
				4.0	2.30865
				5.0	2.07359
				6.0	1.90104
				7.0	1.76685
				8.0	1.65857
				9.0	1.56896
				10.0	1.49348
				15.0	1.24592
			0.160	2.0	3.39474
				3.0	2.74312
				4.0	2.37534
				5.0	2.12790
				6.0	1.94525
				7.0	1.80282
				8.0	1.68781
				9.0	1.59273
				10.0	1.51278
				15.0	1.25268
		0.020	0.080	2.0	3.36307
				3.0	2.68889
				4.0	2.32781
				5.0	2.09462
				6.0	1.92732
				7.0	1.79914
				8.0	1.69652
				9.0	1.61182
				10.0	1.54034
				15.0	1.30109
			0.100	2.0	3.49128
				3.0	2.79938
				4.0	2.42299
				5.0	2.17657
				6.0	1.99786
				7.0	1.85982
				8.0	1.74870
				9.0	1.65667
				10.0	1.57887
				15.0	1.31901

TABLE OF FACTORS

N = 1.0

V	C4	LIFE	VALUE	K	V	C4	LIFE	VALUE
0.020	0.120	2.0	3.61447	0.700	0.030	0.160	2.0	4.43655
		3.0	2.90344				3.0	3.55151
		4.0	2.51085				4.0	3.04741
		5.0	2.25073				5.0	2.70584
		6.0	2.06041				6.0	2.45233
		7.0	1.91257				7.0	2.25381
		8.0	1.79316				8.0	2.09293
		9.0	1.69413				9.0	1.95949
		10.0	1.61042				10.0	1.84693
		15.0	1.33229				15.0	1.47710
	0.140	2.0	3.73283	0.800	0.010	0.080	2.0	2.55580
		3.0	3.00144				3.0	2.14383
		4.0	2.59196				4.0	1.91586
		5.0	2.31782				5.0	1.76385
		6.0	2.11590				6.0	1.65156
		7.0	1.95843				7.0	1.56329
		8.0	1.83105				8.0	1.49105
		9.0	1.72542				9.0	1.43030
		10.0	1.63625				10.0	1.37822
		15.0	1.34213				15.0	1.19865
	0.160	2.0	3.84654			0.100	2.0	2.68512
		3.0	3.09373				3.0	2.25580
		4.0	2.66683				4.0	2.01274
		5.0	2.37853				5.0	1.84764
		6.0	2.16510				6.0	1.72398
		7.0	1.99829				7.0	1.62585
		8.0	1.86333				8.0	1.54506
		9.0	1.75155				9.0	1.47691
		10.0	1.65740				10.0	1.41841
		15.0	1.34941				15.0	1.21766
0.030	0.080	2.0	3.87891			0.120	2.0	2.80937
		3.0	3.08677				3.0	2.36125
		4.0	2.66002				4.0	2.10218
		5.0	2.38286				5.0	1.92345
		6.0	2.18301				6.0	1.78821
		7.0	2.02919				7.0	1.68024
		8.0	1.90556				8.0	1.59109
		9.0	1.80317				9.0	1.51584
		10.0	1.71649				10.0	1.45132
		15.0	1.42420				15.0	1.23175
	0.100	2.0	4.02680			0.140	2.0	2.92874
		3.0	3.21361				3.0	2.46055
		4.0	2.76878				4.0	2.18474
		5.0	2.47609				5.0	1.99205
		6.0	2.26290				6.0	1.84518
		7.0	2.09762				7.0	1.72752
		8.0	1.96417				8.0	1.63031
		9.0	1.85334				9.0	1.54835
		10.0	1.75943				10.0	1.47826
		15.0	1.44382				15.0	1.24219
	0.120	2.0	4.16888			0.160	2.0	3.04344
		3.0	3.33306				3.0	2.55407
		4.0	2.86918				4.0	2.26096
		5.0	2.56045				5.0	2.05413
		6.0	2.33375				6.0	1.89571
		7.0	2.15712				7.0	1.76863
		8.0	2.01411				8.0	1.66374
		9.0	1.89525				9.0	1.57551
		10.0	1.79459				10.0	1.50032
		15.0	1.45836				15.0	1.24992
	0.140	2.0	4.30539		0.020	0.080	2.0	2.89594
		3.0	3.44556				3.0	2.41784
		4.0	2.96186				4.0	2.15096
		5.0	2.63678				5.0	1.97160
		6.0	2.39660				6.0	1.83822
		7.0	2.20884				7.0	1.73279
		8.0	2.05667				8.0	1.64611
		9.0	1.93025				9.0	1.57293
		10.0	1.82337				10.0	1.50997
		15.0	1.46913				15.0	1.29121

N = 1.0

V	FACTOR F6 C4	K 0.800 LIFE	VALUE		C4	LIFE	VALUE
0.020	0.100	2.0	3.04247		0.140	2.0	3.82754
		3.0	2.54412			3.0	3.18566
		4.0	2.25973			4.0	2.80289
		5.0	2.06526			5.0	2.53310
		6.0	1.91883			6.0	2.32618
		7.0	1.80213			7.0	2.15967
		8.0	1.70574			8.0	2.02163
		9.0	1.62418			9.0	1.90489
		10.0	1.55401			10.0	1.80480
		15.0	1.31169			15.0	1.46473
	0.120	2.0	3.18326		0.160	2.0	3.97743
		3.0	2.66304			3.0	3.30675
		4.0	2.36015			4.0	2.90067
		5.0	2.15000			5.0	2.61203
		6.0	1.99032			6.0	2.38998
		7.0	1.86242			7.0	2.21106
		8.0	1.75655			8.0	2.06307
		9.0	1.66699			9.0	1.93831
		10.0	1.59007			10.0	1.83173
		15.0	1.32687			15.0	1.47385
	0.140	2.0	3.31852				
		3.0	2.77504				
		4.0	2.45284				
		5.0	2.22668				
		6.0	2.05373				
		7.0	1.91483				
		8.0	1.79985				
		9.0	1.70275				
		10.0	1.61958				
		15.0	1.33811				

FACTOR		T4		VALUE
E10		0.000		1.00000

WHEN T4 IS GREATER THAN 0, ALL VALUES OF E10 = 0.0

V	FACTOR F6 C4	LIFE	VALUE
	0.160	2.0	3.44848
		3.0	2.88052
		4.0	2.53841
		5.0	2.29607
		6.0	2.10997
		7.0	1.96039
		8.0	1.83675
		9.0	1.73262
		10.0	1.64375
		15.0	1.34644

FACTOR	T3		LIFE	VALUE
E11	2.000		2.0	0.00000
			3.0	2.74592
			4.0	2.16379
			5.0	3.31879
			6.0	2.90049
			7.0	3.57913
			8.0	3.27849
			9.0	3.73367
			10.0	3.51152
			15.0	3.97670

V	FACTOR F6 C4	LIFE	VALUE
0.030	0.080	2.0	3.34014
		3.0	2.77561
		4.0	2.45792
		5.0	2.24291
		6.0	2.08208
		7.0	1.95435
		8.0	1.84894
		9.0	1.75966
		10.0	1.68265
		15.0	1.41339

T3		LIFE	VALUE
3.000		2.0	0.00000
		3.0	0.00000
		4.0	1.96708
		5.0	1.65189
		6.0	1.44368
		7.0	2.27074
		8.0	2.08001
		9.0	1.93381
		10.0	2.40496
		15.0	2.33359

V	FACTOR F6 C4	LIFE	VALUE
	0.100	2.0	3.50914
		3.0	2.92057
		4.0	2.58222
		5.0	2.34946
		6.0	2.17338
		7.0	2.03257
		8.0	1.91592
		9.0	1.81700
		10.0	1.73172
		15.0	1.43581
	0.120	2.0	3.67152
		3.0	3.05709
		4.0	2.69697
		5.0	2.44586
		6.0	2.25436
		7.0	2.10056
		8.0	1.97299
		9.0	1.86489
		10.0	1.77190
		15.0	1.45242

T3		LIFE	VALUE
4.000		2.0	0.00000
		3.0	0.00000
		4.0	0.00000
		5.0	1.50172
		6.0	1.31244
		7.0	1.17872
		8.0	1.07971
		9.0	1.68944
		10.0	1.58893
		15.0	1.66434

FACTOR		VALUE
D1		0.47537

TABLE OF FACTORS

FACTOR	N = 1.0 V			LIFE	VALUE
D2	0.010			3.0	3.58377
				4.0	2.61573
				5.0	2.06310
				6.0	1.70608
				7.0	1.45657
				8.0	1.27245
				9.0	1.13106
				10.0	1.01910
				15.0	0.68981
	0.020			3.0	4.04182
				4.0	2.93671
				5.0	2.30610
				6.0	1.89891
				7.0	1.61450
				8.0	1.40478
				9.0	1.24384
				10.0	1.11653
				15.0	0.74307
	0.030			3.0	4.63989
				4.0	3.35582
				5.0	2.62344
				6.0	2.15082
				7.0	1.82094
				8.0	1.57787
				9.0	1.39151
				10.0	1.24421
				15.0	0.81339

FACTOR	C3				VALUE
D5	0.080				0.05500
	0.100				0.07262
	0.120				0.08988
	0.140				0.10680
	0.160				0.12339

FACTOR	K	V	C4	LIFE	VALUE
D6	0.700	0.010	0.080	2.0	1.98219
				3.0	1.42770
				4.0	1.14331
				5.0	0.96750
				6.0	0.84637
				7.0	0.75682
				8.0	0.68724
				9.0	0.63119
				10.0	0.58476
				15.0	0.43282
			0.100	2.0	2.10945
				3.0	1.54475
				4.0	1.25103
				5.0	1.06667
				6.0	0.93773
				7.0	0.84102
				8.0	0.76489
				9.0	0.70282
				10.0	0.65088
				15.0	0.47744
			0.120	2.0	2.23173
				3.0	1.65499
				4.0	1.35046
				5.0	1.15641
				6.0	1.01876
				7.0	0.91422
				8.0	0.83105
				9.0	0.76266
				10.0	0.70502
				15.0	0.51050

CONSTRUCTION EQUIPMENT POLICY

$N = 1.0$

V	FACTOR D6 C4	K 0.700 LIFE	VALUE	K	V	C4	LIFE	VALUE
0.010	0.140	2.0	2.34921	0.030		0.080	2.0	2.59049
		3.0	1.75880				3.0	1.84844
		4.0	1.44225				4.0	1.46680
		5.0	1.23760				5.0	1.23027
		6.0	1.09062				6.0	1.06701
		7.0	0.97786				7.0	0.94614
		8.0	0.88743				8.0	0.85220
		9.0	0.81264				9.0	0.77653
		10.0	0.74935				10.0	0.71393
		15.0	0.53499				15.0	0.51035
	0.160	2.0	2.46208			0.100	2.0	2.75682
		3.0	1.85657				3.0	1.99999
		4.0	1.52698				4.0	1.60499
		5.0	1.31107				5.0	1.35638
		6.0	1.15436				6.0	1.18218
		7.0	1.03318				7.0	1.05140
		8.0	0.93548				8.0	0.94848
		9.0	0.85438				9.0	0.86466
		10.0	0.78564				10.0	0.79466
		15.0	0.55314				15.0	0.56298
0.020	0.080	2.0	2.24599			0.120	2.0	2.91662
		3.0	1.61018				3.0	2.14271
		4.0	1.28361				4.0	1.73255
		5.0	1.08145				5.0	1.47049
		6.0	0.94203				6.0	1.28433
		7.0	0.83888				7.0	1.14292
		8.0	0.75871				8.0	1.03052
		9.0	0.69413				9.0	0.93828
		10.0	0.64066				10.0	0.86075
		15.0	0.46624				15.0	0.60196
	0.100	2.0	2.39020			0.140	2.0	3.07015
		3.0	1.74219				3.0	2.27712
		4.0	1.40454				4.0	1.85031
		5.0	1.19231				5.0	1.57374
		6.0	1.04372				6.0	1.37493
		7.0	0.93221				7.0	1.22248
		8.0	0.84443				8.0	1.10044
		9.0	0.77291				9.0	0.99976
		10.0	0.71311				10.0	0.91487
		15.0	0.51431				15.0	0.63084
	0.120	2.0	2.52874			0.160	2.0	3.21767
		3.0	1.86652				3.0	2.40370
		4.0	1.51618				4.0	1.95902
		5.0	1.29261				5.0	1.66716
		6.0	1.13390				6.0	1.45528
		7.0	1.01335				7.0	1.29164
		8.0	0.91747				8.0	1.16002
		9.0	0.83871				9.0	1.05112
		10.0	0.77242				10.0	0.95917
		15.0	0.54992				15.0	0.65223
	0.140	2.0	2.66186	0.800	0.010	0.080	2.0	1.51851
		3.0	1.98360				3.0	1.14056
		4.0	1.61923				4.0	0.94317
		5.0	1.38337				5.0	0.81862
		6.0	1.21389				6.0	0.73096
		7.0	1.08388				7.0	0.66474
		8.0	0.97972				8.0	0.61222
		9.0	0.89367				9.0	0.56906
		10.0	0.82098				10.0	0.53265
		15.0	0.57631				15.0	0.40822
	0.160	2.0	2.78976			0.100	2.0	1.66396
		3.0	2.09387				3.0	1.27434
		4.0	1.71436				4.0	1.06627
		5.0	1.46550				5.0	0.93196
		6.0	1.28483				6.0	0.83537
		7.0	1.14521				7.0	0.76097
		8.0	1.03276				8.0	0.70096
		9.0	0.93958				9.0	0.65093
		10.0	0.86074				10.0	0.60822
		15.0	0.59585				15.0	0.45922

TABLE OF FACTORS

$N = 1.0$

V	C4	LIFE	VALUE	V	C4	LIFE	VALUE
0.010	0.120	2.0	1.80370	0.020	0.160	2.0	2.34206
		3.0	1.40032			3.0	1.83913
		4.0	1.17991			4.0	1.55119
		5.0	1.03452			5.0	1.35394
		6.0	0.92797			6.0	1.20534
		7.0	0.84463			7.0	1.08691
		8.0	0.77658			8.0	0.98909
		9.0	0.71931			9.0	0.90632
		10.0	0.67009			10.0	0.83509
		15.0	0.49700			15.0	0.58787
	0.140	2.0	1.93797	0.030	0.080	2.0	1.98453
		3.0	1.51897			3.0	1.47668
		4.0	1.28481			4.0	1.21003
		5.0	1.12731			5.0	1.04096
		6.0	1.01010			6.0	0.92150
		7.0	0.91736			7.0	0.83103
		8.0	0.84101			8.0	0.75917
		9.0	0.77643			9.0	0.70010
		10.0	0.72075			10.0	0.65030
		15.0	0.52499			15.0	0.48135
	0.160	2.0	2.06697		0.100	2.0	2.17461
		3.0	1.63070			3.0	1.64988
		4.0	1.38164			4.0	1.36796
		5.0	1.21128			5.0	1.18508
		6.0	1.08294			6.0	1.05313
		7.0	0.98059			7.0	0.95134
		8.0	0.89592			8.0	0.86921
		9.0	0.82414			9.0	0.80082
		10.0	0.76222			10.0	0.74257
		15.0	0.54573			15.0	0.54149
0.020	0.080	2.0	1.72061		0.120	2.0	2.35724
		3.0	1.28634			3.0	1.81299
		4.0	1.05891			4.0	1.51375
		5.0	0.91504			5.0	1.31549
		6.0	0.81357			6.0	1.16987
		7.0	0.73682			7.0	1.05592
		8.0	0.67589			8.0	0.96297
		9.0	0.62581			9.0	0.88495
		10.0	0.58357			10.0	0.81811
		15.0	0.43974			15.0	0.58604
	0.100	2.0	1.88541		0.140	2.0	2.53271
		3.0	1.43721			3.0	1.96660
		4.0	1.19712			4.0	1.64833
		5.0	1.04173			5.0	1.43349
		6.0	0.92978			6.0	1.27341
		7.0	0.84348			7.0	1.14684
		8.0	0.77385			8.0	1.04288
		9.0	0.71584			9.0	0.95522
		10.0	0.66636			10.0	0.87995
		15.0	0.49468			15.0	0.61905
	0.120	2.0	2.04376		0.160	2.0	2.70130
		3.0	1.57930			3.0	2.11126
		4.0	1.32470			4.0	1.77256
		5.0	1.15636			5.0	1.54026
		6.0	1.03285			6.0	1.36524
		7.0	0.93621			7.0	1.22589
		8.0	0.85733			8.0	1.11096
		9.0	0.79104			9.0	1.01392
		10.0	0.73415			10.0	0.93059
		15.0	0.53538			15.0	0.64350
	0.140	2.0	2.19589				
		3.0	1.71311				
		4.0	1.44247				
		5.0	1.26009				
		6.0	1.12426				
		7.0	1.01683				
		8.0	0.92847				
		9.0	0.85386				
		10.0	0.78965				
		15.0	0.56553				

FACTOR							VALUE
T1							0.09445

			N = 1.0	
FACTOR	V		LIFE	VALUE
T2	0.010		2.0	1.04558
			3.0	1.03130
			4.0	1.01773
			5.0	0.84735
			6.0	0.73445
			7.0	0.65439
			8.0	0.59485
			9.0	0.54899
			10.0	0.51269
			15.0	0.40761
	0.020		2.0	1.18473
			3.0	1.16312
			4.0	1.14261
			5.0	0.94715
			6.0	0.81746
			7.0	0.72535
			8.0	0.65671
			9.0	0.60373
			10.0	0.56170
			15.0	0.43908
	0.030		2.0	1.36646
			3.0	1.33522
			4.0	1.30568
			5.0	1.07749
			6.0	0.92591
			7.0	0.81809
			8.0	0.73763
			9.0	0.67540
			10.0	0.62594
			15.0	0.48063

FACTOR	C3	VALUE
G1	0.080	0.44176
	0.100	0.42312
	0.120	0.40484
	0.140	0.38692
	0.160	0.36936

FACTOR	K	V	C4	LIFE	VALUE
G2	0.700	0.010	0.080	3.0	1.57575
				4.0	0.97750
				5.0	0.65491
				6.0	0.45966
				7.0	0.33260
				8.0	0.24577
				9.0	0.18435
				10.0	0.13979
				15.0	0.03783
			0.100	3.0	1.47778
				4.0	0.89272
				5.0	0.58159
				6.0	0.39629
				7.0	0.27785
				8.0	0.19851
				9.0	0.14357
				10.0	0.10462
				15.0	0.02119

TABLE OF FACTORS

V	C4	LIFE	VALUE	N = 1.0 K	V	C4	LIFE	VALUE
	0.120	3.0	1.38552		0.030	0.080	3.0	2.04012
		4.0	0.81446				4.0	1.25406
		5.0	0.51525				5.0	0.83278
		6.0	0.34009				6.0	0.57949
		7.0	0.23026				7.0	0.41580
		8.0	0.15823				8.0	0.30476
		9.0	0.10951				9.0	0.22680
		10.0	0.07583				10.0	0.17067
		15.0	0.00887				15.0	0.04461
	0.140	3.0	1.29862			0.100	3.0	1.91328
		4.0	0.74222				4.0	1.14530
		5.0	0.45522				5.0	0.73955
		6.0	0.29024				6.0	0.49960
		7.0	0.18989				7.0	0.34736
		8.0	0.12391				8.0	0.24615
		9.0	0.08105				9.0	0.17663
		10.0	0.05225				10.0	0.12773
		15.0	-0.00027				15.0	0.02499
	0.160	3.0	1.21679			0.120	3.0	1.79382
		4.0	0.67553				4.0	1.04490
		5.0	0.40091				5.0	0.65519
		6.0	0.24603				6.0	0.42874
		7.0	0.15292				7.0	0.28787
		8.0	0.09467				8.0	0.19621
		9.0	0.05729				9.0	0.13472
		10.0	0.03295				10.0	0.09258
		15.0	-0.00703				15.0	0.01045
0.020	0.080	3.0	1.77715			0.140	3.0	1.68132
		4.0	1.09745				4.0	0.95222
		5.0	0.73204				5.0	0.57886
		6.0	0.51161				6.0	0.36590
		7.0	0.36866				7.0	0.23614
		8.0	0.27133				8.0	0.15365
		9.0	0.20273				9.0	0.09972
		10.0	0.15315				10.0	0.06379
		15.0	0.04075				15.0	-0.00031
	0.100	3.0	1.66666			0.160	3.0	1.57538
		4.0	1.00227				4.0	0.86667
		5.0	0.65009				5.0	0.50980
		6.0	0.44108				6.0	0.31016
		7.0	0.30798				7.0	0.19118
		8.0	0.21915				8.0	0.11739
		9.0	0.15789				9.0	0.07048
		10.0	0.11462				10.0	0.04023
		15.0	0.02283				15.0	-0.00829
	0.120	3.0	1.56260	0.800	0.010	0.080	3.0	1.81608
		4.0	0.91441				4.0	1.13502
		5.0	0.57594				5.0	0.76497
		6.0	0.37852				6.0	0.53972
		7.0	0.25523				7.0	0.39246
		8.0	0.17469				8.0	0.29143
		9.0	0.12043				9.0	0.21972
		10.0	0.08308				10.0	0.16751
		15.0	0.00955				15.0	0.04700
	0.140	3.0	1.46461			0.100	3.0	1.70412
		4.0	0.83330				4.0	1.03813
		5.0	0.50884				5.0	0.68118
		6.0	0.32304				6.0	0.46730
		7.0	0.20937				7.0	0.32989
		8.0	0.13680				8.0	0.23742
		9.0	0.08914				9.0	0.17311
		10.0	0.05725				10.0	0.12731
		15.0	-0.00029				15.0	0.02799
	0.160	3.0	1.37231					
		4.0	0.75843					
		5.0	0.44813					
		6.0	0.27383					
		7.0	0.16950					
		8.0	0.10451					
		9.0	0.06300					
		10.0	0.03610					
		15.0	-0.00758					

N = 1.0

FACTOR G2   K 0.800

V	C4	LIFE	VALUE	V	C4	LIFE	VALUE
0.010	0.120	3.0	1.59867	0.030	0.080	3.0	2.35128
		4.0	0.94869			4.0	1.45615
		5.0	0.60536			5.0	0.97273
		6.0	0.40306			6.0	0.68041
		7.0	0.27550			7.0	0.49063
		8.0	0.19139			8.0	0.36138
		9.0	0.13418			9.0	0.27031
		10.0	0.09441			10.0	0.20451
		15.0	0.01390			15.0	0.05543
	0.140	3.0	1.49937		0.100	3.0	2.20631
		4.0	0.86613			4.0	1.33185
		5.0	0.53676			5.0	0.86619
		6.0	0.34609			6.0	0.58911
		7.0	0.22822			7.0	0.41242
		8.0	0.15217			8.0	0.29440
		9.0	0.10166			9.0	0.21297
		10.0	0.06746			10.0	0.15544
		15.0	0.00346			15.0	0.03300
	0.160	3.0	1.40584		0.120	3.0	2.06979
		4.0	0.78992			4.0	1.21711
		5.0	0.47469			5.0	0.76978
		6.0	0.29557			6.0	0.50813
		7.0	0.18712			7.0	0.34442
		8.0	0.11875			8.0	0.23733
		9.0	0.07450			9.0	0.16508
		10.0	0.04540			10.0	0.11526
		15.0	-0.00427			15.0	0.01639
0.020	0.080	3.0	2.04820		0.140	3.0	1.94122
		4.0	1.27430			4.0	1.11119
		5.0	0.85507			5.0	0.68254
		6.0	0.60072			6.0	0.43631
		7.0	0.43501			7.0	0.28531
		8.0	0.32174			8.0	0.18869
		9.0	0.24163			9.0	0.12507
		10.0	0.18352			10.0	0.08237
		15.0	0.05063			15.0	0.00408
	0.100	3.0	1.92193		0.160	3.0	1.82014
		4.0	1.16552			4.0	1.01341
		5.0	0.76141			5.0	0.60361
		6.0	0.52011			6.0	0.37261
		7.0	0.36566			7.0	0.23392
		8.0	0.26211			8.0	0.14725
		9.0	0.19037			9.0	0.09166
		10.0	0.13949			10.0	0.05543
		15.0	0.03015			15.0	-0.00503
	0.120	3.0	1.80300				
		4.0	1.06511				
		5.0	0.67666				
		6.0	0.44862				
		7.0	0.30538				
		8.0	0.21129				
		9.0	0.14756				
		10.0	0.10343				
		15.0	0.01497				
	0.140	3.0	1.69101				
		4.0	0.97242				
		5.0	0.59998				
		6.0	0.38521				
		7.0	0.25297				
		8.0	0.16799				
		9.0	0.11180				
		10.0	0.07391				
		15.0	0.00373				
	0.160	3.0	1.58553				
		4.0	0.88685				
		5.0	0.53060				
		6.0	0.32897				
		7.0	0.20740				
		8.0	0.13109				
		9.0	0.08193				
		10.0	0.04974				
		15.0	-0.00460				

FACTOR	V	VALUE
TC11	0.010	0.91823
	0.020	0.92746
	0.030	0.93678

FACTOR	C3	VALUE
TC12	0.080	0.83920
	0.100	0.82258
	0.120	0.80629
	0.140	0.79033
	0.160	0.77468

TABLE OF FACTORS

N = 1.0

FACTOR	K	V	C4		
IC2	0.700	0.010	0.080		
				3.0	1.46594
				4.0	1.15515
				5.0	0.95568
				6.0	0.81339
				7.0	0.70492
				8.0	0.61849
				9.0	0.54744
				10.0	0.48771
				15.0	0.28959
			0.100		
				3.0	1.56391
				4.0	1.23993
				5.0	1.02900
				6.0	0.87676
				7.0	0.75966
				8.0	0.66575
				9.0	0.58822
				10.0	0.52288
				15.0	0.30623
			0.120		
				3.0	1.65618
				4.0	1.31819
				5.0	1.09534
				6.0	0.93296
				7.0	0.80725
				8.0	0.70603
				9.0	0.62228
				10.0	0.55167
				15.0	0.31856
			0.140		
				3.0	1.74307
				4.0	1.39043
				5.0	1.15536
				6.0	0.98281
				7.0	0.84863
				8.0	0.74034
				9.0	0.65074
				10.0	0.57525
				15.0	0.32769
			0.160		
				3.0	1.82490
				4.0	1.45712
				5.0	1.20968
				6.0	1.02702
				7.0	0.88460
				8.0	0.76959
				9.0	0.67450
				10.0	0.59455
				15.0	0.33446
		0.020	0.080		
				3.0	1.76144
				4.0	1.40036
				5.0	1.16717
				6.0	0.99987
				7.0	0.87169
				8.0	0.76907
				9.0	0.68436
				10.0	0.61289
				15.0	0.37363
			0.100		
				3.0	1.87193
				4.0	1.49553
				5.0	1.24912
				6.0	1.07040
				7.0	0.93236
				8.0	0.82124
				9.0	0.72921
				10.0	0.65142
				15.0	0.39156

$N = 1.0$

V	FACTOR IC2 C4	K	LIFE	VALUE	K	V	C4	LIFE	VALUE
0.020	0.120	0.700			0.700	0.030	0.160		
			3.0	1.97599				3.0	2.61474
			4.0	1.58340				4.0	2.11064
			5.0	1.32327				5.0	1.76907
			6.0	1.13296				6.0	1.51556
			7.0	0.98511				7.0	1.31703
			8.0	0.86571				8.0	1.15616
			9.0	0.76667				9.0	1.02271
			10.0	0.68297				10.0	0.91016
			15.0	0.40484				15.0	0.54033
	0.140				0.800	0.010	0.080		
			3.0	2.07399				3.0	1.22561
			4.0	1.66450				4.0	0.99763
			5.0	1.39037				5.0	0.84562
			6.0	1.18844				6.0	0.73333
			7.0	1.03097				7.0	0.64506
			8.0	0.90359				8.0	0.57283
			9.0	0.79796				9.0	0.51207
			10.0	0.70879				10.0	0.45999
			15.0	0.41467				15.0	0.28042
	0.160						0.100		
			3.0	2.16628				3.0	1.33758
			4.0	1.73937				4.0	1.09452
			5.0	1.45108				5.0	0.92941
			6.0	1.23765				6.0	0.80575
			7.0	1.07084				7.0	0.70762
			8.0	0.93588				8.0	0.62684
			9.0	0.82410				9.0	0.55868
			10.0	0.72994				10.0	0.50019
			15.0	0.42196				15.0	0.29944
0.030	0.080						0.120		
			3.0	2.14999				3.0	1.44302
			4.0	1.72324				4.0	1.18396
			5.0	1.44609				5.0	1.00523
			6.0	1.24623				6.0	0.86999
			7.0	1.09241				7.0	0.76201
			8.0	0.96879				8.0	0.67287
			9.0	0.86639				9.0	0.59761
			10.0	0.77972				10.0	0.53310
			15.0	0.48742				15.0	0.31353
	0.100						0.140		
			3.0	2.27684				3.0	1.54233
			4.0	1.83201				4.0	1.26652
			5.0	1.53932				5.0	1.07383
			6.0	1.32612				6.0	0.92696
			7.0	1.16085				7.0	0.80930
			8.0	1.02739				8.0	0.71209
			9.0	0.91656				9.0	0.63013
			10.0	0.82266				10.0	0.56004
			15.0	0.50705				15.0	0.32396
	0.120						0.160		
			3.0	2.39629				3.0	1.63585
			4.0	1.93241				4.0	1.34273
			5.0	1.62367				5.0	1.13590
			6.0	1.39698				6.0	0.97748
			7.0	1.22034				7.0	0.85040
			8.0	1.07733				8.0	0.74551
			9.0	0.95847				9.0	0.65729
			10.0	0.85781				10.0	0.58210
			15.0	0.52158				15.0	0.33169
	0.140					0.020	0.080		
			3.0	2.50879				3.0	1.49039
			4.0	2.02509				4.0	1.22350
			5.0	1.70000				5.0	1.04414
			6.0	1.45982				6.0	0.91076
			7.0	1.27207				7.0	0.80533
			8.0	1.11989				8.0	0.71865
			9.0	0.99348				9.0	0.64547
			10.0	0.88659				10.0	0.58252
			15.0	0.53235				15.0	0.36375

TABLE OF FACTORS

$N = 1.0$

V	C4	LIFE	VALUE		C4	LIFE	VALUE
0.020	0.100				0.140		
		3.0	1.61667			3.0	2.24889
		4.0	1.33228			4.0	1.86612
		5.0	1.13780			5.0	1.59632
		6.0	0.99137			6.0	1.38941
		7.0	0.87468			7.0	1.22290
		8.0	0.77828			8.0	1.08485
		9.0	0.69673			9.0	0.96812
		10.0	0.62656			10.0	0.86802
		15.0	0.38424			15.0	0.52795
	0.120				0.160		
		3.0	1.73559			3.0	2.36997
		4.0	1.43269			4.0	1.96389
		5.0	1.22255			5.0	1.67525
		6.0	1.06287			6.0	1.45310
		7.0	0.93497			7.0	1.27429
		8.0	0.82910			8.0	1.12630
		9.0	0.73954			9.0	1.00153
		10.0	0.66261			10.0	0.89495
		15.0	0.39941			15.0	0.53707
	0.140						
		3.0	1.84759				
		4.0	1.52539				
		5.0	1.29923				
		6.0	1.12627				
		7.0	0.98738				
		8.0	0.87240				
		9.0	0.77530				
		10.0	0.69213				
		15.0	0.41066				
	0.160						
		3.0	1.95306				
		4.0	1.61095				
		5.0	1.36861				
		6.0	1.18251				
		7.0	1.03294				
		8.0	0.90930				
		9.0	0.80516				
		10.0	0.71630				
		15.0	0.41898				
0.030	0.080						
		3.0	1.83884				
		4.0	1.52115				
		5.0	1.30613				
		6.0	1.14531				
		7.0	1.01758				
		8.0	0.91216				
		9.0	0.82288				
		10.0	0.74588				
		15.0	0.47661				
	0.100						
		3.0	1.98380				
		4.0	1.64545				
		5.0	1.41268				
		6.0	1.23661				
		7.0	1.09579				
		8.0	0.97914				
		9.0	0.88022				
		10.0	0.79495				
		15.0	0.49903				
	0.120						
		3.0	2.12032				
		4.0	1.76020				
		5.0	1.50909				
		6.0	1.31758				
		7.0	1.16378				
		8.0	1.03622				
		9.0	0.92812				
		10.0	0.83513				
		15.0	0.51565				

ALL FACTORS FOR N=2.0 ARE COMPUTED TO HAVE THE
FOLLOWING VALUES AT 10% RATE OF RETURN

$N = 2.0$

FACTOR	S1	VALUE
S1	0.020	1.78613
	0.040	1.75222
	0.060	1.71919
	0.080	1.68700
	0.100	1.65564

$N = 2.0$

FACTOR	S2		LIFE	VALUE	G	S2	LIFE	VALUE
P2	0.020		2.0	8.50541		0.060	2.0	6.62807
			3.0	8.42815			3.0	6.51143
			4.0	8.35450			4.0	6.40252
			5.0	8.28435			5.0	6.30095
			6.0	8.21759			6.0	6.20632
			7.0	8.15412			7.0	6.11828
			8.0	8.09383			8.0	6.03646
			9.0	8.03661			9.0	5.96050
			10.0	7.98235			10.0	5.89008
			15.0	7.75183			15.0	5.60991
	0.040		2.0	8.34394		0.080	2.0	6.50397
			3.0	8.19442			3.0	6.33461
			4.0	8.05367			4.0	6.17846
			5.0	7.92130			5.0	6.03461
			6.0	7.79689			6.0	5.90221
			7.0	7.68006			7.0	5.78049
			8.0	7.57042			8.0	5.66868
			9.0	7.46760			9.0	5.56606
			10.0	7.37126			10.0	5.47197
			15.0	6.97515			15.0	5.10880
	0.060		2.0	8.18664		0.100	2.0	6.38305
			3.0	7.96953			3.0	6.16442
			4.0	7.76776			4.0	5.96533
			5.0	7.58033			5.0	5.78416
			6.0	7.40633			6.0	5.61939
			7.0	7.24487			7.0	5.46964
			8.0	7.09513			8.0	5.33363
			9.0	6.95632			9.0	5.21017
			10.0	6.82773			10.0	5.09819
			15.0	6.31516			15.0	4.67801
	0.080		2.0	8.03337	0.040	0.020	2.0	5.74731
			3.0	7.75312			3.0	5.79904
			4.0	7.49591			4.0	5.84881
			5.0	7.25991			5.0	5.89662
			6.0	7.04342			6.0	5.94243
			7.0	6.84488			7.0	5.98624
			8.0	6.66284			8.0	6.02803
			9.0	6.49598			9.0	6.06782
			10.0	6.34306			10.0	6.10563
			15.0	5.75105			15.0	6.26603
	0.100		2.0	7.88400		0.040	2.0	5.63821
			3.0	7.54481			3.0	5.63821
			4.0	7.23734			4.0	5.63821
			5.0	6.95861			5.0	5.63821
			6.0	6.70591			6.0	5.63821
			7.0	6.47679			7.0	5.63821
			8.0	6.26904			8.0	5.63821
			9.0	6.08064			9.0	5.63821
			10.0	5.90977			10.0	5.63821
			15.0	5.26610			15.0	5.63821
						0.060	2.0	5.53192
							3.0	5.48348
							4.0	5.43805
FACTOR	G	S2	LIFE	VALUE			5.0	5.39552
P3	0.020	0.020	2.0	6.88615			6.0	5.35578
			3.0	6.88614			7.0	5.31872
			4.0	6.88614			8.0	5.28423
			5.0	6.88614			9.0	5.25218
			6.0	6.88614			10.0	5.22247
			7.0	6.88614			15.0	5.10472
			8.0	6.88615				
			9.0	6.88614		0.080	2.0	5.42835
			10.0	6.88614			3.0	5.33458
			15.0	6.88614			4.0	5.24773
							5.0	5.16745
							6.0	5.09335
		0.040	2.0	6.75542			7.0	5.02507
			3.0	6.69517			8.0	4.96227
			4.0	6.63819			9.0	4.90461
			5.0	6.58437			10.0	4.85175
			6.0	6.53361			15.0	4.64874
			7.0	6.48580				
			8.0	6.44083				
			9.0	6.39859				
			10.0	6.35897				
			15.0	6.19620				

TABLE OF FACTORS

$N = 2.0$

G	S2	LIFE	VALUE		S2	LIFE	VALUE
	0.100	2.0	5.32742		0.040	2.0	4.17248
		3.0	5.19125			3.0	4.24583
		4.0	5.06672			4.0	4.31609
		5.0	4.95299			5.0	4.38315
		6.0	4.84978			6.0	4.44692
		7.0	4.75485			7.0	4.50734
		8.0	4.66898			8.0	4.56438
		9.0	4.59102			9.0	4.61804
		10.0	4.52033			10.0	4.66835
		15.0	4.25674			15.0	4.87223
0.060	0.020	2.0	4.90331		0.060	2.0	4.09382
		3.0	4.99114			3.0	4.12931
		4.0	5.07604			4.0	4.16286
		5.0	5.15787			5.0	4.19448
		6.0	5.23650			6.0	4.22416
		7.0	5.31186			7.0	4.25193
		8.0	5.38386			8.0	4.27781
		9.0	5.45247			9.0	4.30186
		10.0	5.51766			10.0	4.32413
		15.0	5.79323			15.0	4.41121
	0.040	2.0	4.81023		0.080	2.0	4.01718
		3.0	4.85272			3.0	4.01718
		4.0	4.89326			4.0	4.01718
		5.0	4.93183			5.0	4.01717
		6.0	4.96842			6.0	4.01718
		7.0	5.00304			7.0	4.01718
		8.0	5.03570			8.0	4.01718
		9.0	5.06642			9.0	4.01718
		10.0	5.09525			10.0	4.01718
		15.0	5.21278			15.0	4.01718
	0.060	2.0	4.71955		0.100	2.0	3.94249
		3.0	4.71954			3.0	3.90925
		4.0	4.71954			4.0	3.87861
		5.0	4.71954			5.0	3.85046
		6.0	4.71954			6.0	3.82468
		7.0	4.71955			7.0	3.80115
		8.0	4.71955			8.0	3.77974
		9.0	4.71955			9.0	3.76033
		10.0	4.71955			10.0	3.74277
		15.0	4.71955			15.0	3.67843
	0.080	2.0	4.63118				
		3.0	4.59139				
		4.0	4.55438				
		5.0	4.52005	FACTOR			VALUE
		6.0	4.48828				
		7.0	4.45898	1			1.82094
		8.0	4.43200				
		9.0	4.40722				
		10.0	4.38453				
		15.0	4.29797				
				FACTOR	W1		VALUE
	0.100	2.0	4.54508				
		3.0	4.46803	S2	0.020		1.78613
		4.0	4.39727				
		5.0	4.33246		0.040		1.75222
		6.0	4.27321				
		7.0	4.21919		0.060		1.71919
		8.0	4.17004				
		9.0	4.12543		0.080		1.68700
		10.0	4.08503				
		15.0	3.93555		0.100		1.65564
0.080	0.020	2.0	4.25323		0.120		1.62507
		3.0	4.36694				
		4.0	4.47731				
		5.0	4.58403				
		6.0	4.68686			LIFE	VALUE
		7.0	4.78556	FACTOR			
		8.0	4.87995			2.0	7.76707
		9.0	4.96992	E3		3.0	7.80275
		10.0	5.05537			4.0	7.83729
		15.0	5.41475			5.0	7.87069
						6.0	7.90295
						7.0	7.93406
						8.0	7.96402
						9.0	7.99284
						10.0	8.02053
						15.0	8.14247

CONSTRUCTION EQUIPMENT POLICY

N = 2.0

FACTOR	LIFE	VALUE	FACTOR	C3	LIFE	VALUE
E4	2.0	7.29543	F6	0.080		0.29575
	3.0	7.36298				
	4.0	7.42854		0.100		0.32336
	5.0	7.49207				
	6.0	7.55354		0.120		0.34989
	7.0	7.61292				
	8.0	7.67019		0.140		0.37539
	9.0	7.72534				
	10.0	7.77835		0.160		0.39988
	15.0	8.01190				

FACTOR	W2	LIFE	VALUE	FACTOR	K	V	C4	LIFE	VALUE
E5	0.020	2.0	7.15599	F7	0.700	0.010	0.080	2.0	2.72535
		3.0	7.15664					3.0	2.18921
		4.0	7.15727					4.0	1.90383
		5.0	7.15788					5.0	1.72067
		6.0	7.15845					6.0	1.59002
		7.0	7.15900					7.0	1.49042
		8.0	7.15952					8.0	1.41105
		9.0	7.16001					9.0	1.34582
		10.0	7.16048					10.0	1.29097
		15.0	7.16248					15.0	1.10905
	0.040	2.0	7.02014				0.100	2.0	2.82925
		3.0	6.95817					3.0	2.27917
		4.0	6.89955					4.0	1.98168
		5.0	6.84419					5.0	1.78799
		6.0	6.79197					6.0	1.64820
		7.0	6.74279					7.0	1.54069
		8.0	6.69653					8.0	1.45445
		9.0	6.65307					9.0	1.38326
		10.0	6.61231					10.0	1.32326
		15.0	6.44485					15.0	1.12433
	0.060	2.0	6.88780				0.120	2.0	2.92908
		3.0	6.76721					3.0	2.36389
		4.0	6.65461					4.0	2.05354
		5.0	6.54958					5.0	1.84891
		6.0	6.45175					6.0	1.69981
		7.0	6.36071					7.0	1.58438
		8.0	6.27610					8.0	1.49143
		9.0	6.19756						
		10.0	6.12473						
		15.0	5.83503						
	0.080	2.0	6.75884						
		3.0	6.58345						
		4.0	6.42172						
		5.0	6.27273						
		6.0	6.13561					9.0	1.41454
		7.0	6.00953					10.0	1.34970
		8.0	5.89371					15.0	1.13565
		9.0	5.78742				0.140	2.0	3.02499
		10.0	5.68997					3.0	2.44367
		15.0	5.31381					4.0	2.11987
	0.100	2.0	6.63318					5.0	1.90403
		3.0	6.40657					6.0	1.74558
		4.0	6.20021					7.0	1.62237
		5.0	6.01240					8.0	1.52295
		6.0	5.84160					9.0	1.44067
		7.0	5.68637					10.0	1.37135
		8.0	5.54537					15.0	1.14404
		9.0	5.41739				0.160	2.0	3.11715
		10.0	5.30130					3.0	2.51881
		15.0	4.86573					4.0	2.18110
	0.120	2.0	6.51071					5.0	1.95390
		3.0	6.23628					6.0	1.78618
		4.0	5.98943					7.0	1.65540
		5.0	5.76746					8.0	1.54980
		6.0	5.56792					9.0	1.46249
		7.0	5.38861					10.0	1.38907
		8.0	5.22751					15.0	1.15025
		9.0	5.08283						
		10.0	4.95293						
		15.0	4.47831						

TABLE OF FACTORS 279

$N = 2.0$

V	C4	LIFE	VALUE	K	V	C4	LIFE	VALUE
0.020	0.080	2.0	3.11910			0.120	2.0	3.90531
		3.0	2.49383				3.0	3.12234
		4.0	2.15894				4.0	2.68778
		5.0	1.94267				5.0	2.39857
		6.0	1.78751				6.0	2.18621
		7.0	1.66862				7.0	2.02074
		8.0	1.57345				8.0	1.88677
		9.0	1.49489				9.0	1.77542
		10.0	1.42860				10.0	1.68113
		15.0	1.20670				15.0	1.36616
	0.100	2.0	3.23801			0.140	2.0	4.03319
		3.0	2.59631				3.0	3.22772
		4.0	2.24722				4.0	2.77460
		5.0	2.01868				5.0	2.47007
		6.0	1.85292				6.0	2.24508
		7.0	1.72490				7.0	2.06919
		8.0	1.62184				8.0	1.92664
		9.0	1.53648				9.0	1.80822
		10.0	1.46434				10.0	1.70809
		15.0	1.22332				15.0	1.37624
	0.120	2.0	3.35226			0.160	2.0	4.15606
		3.0	2.69282				3.0	3.32697
		4.0	2.32870				4.0	2.85475
		5.0	2.08745				5.0	2.53477
		6.0	1.91094				6.0	2.29729
		7.0	1.77382				7.0	2.11132
		8.0	1.66308				8.0	1.96061
		9.0	1.57123				9.0	1.83560
		10.0	1.49359				10.0	1.73016
		15.0	1.23564				15.0	1.38372
	0.140	2.0	3.46203	0.800	0.010	0.080	2.0	2.34680
		3.0	2.78370				3.0	1.96853
		4.0	2.40392				4.0	1.75919
		5.0	2.14968				5.0	1.61961
		6.0	1.96240				6.0	1.51651
		7.0	1.81635				7.0	1.43545
		8.0	1.69822				8.0	1.36912
		9.0	1.60025				9.0	1.31334
		10.0	1.51755				10.0	1.26552
		15.0	1.24476				15.0	1.10063
	0.160	2.0	3.56750			0.100	2.0	2.46555
		3.0	2.86930				3.0	2.07134
		4.0	2.47336				4.0	1.84816
		5.0	2.20599				5.0	1.69555
		6.0	2.00804				6.0	1.58301
		7.0	1.85333				7.0	1.49290
		8.0	1.72816				8.0	1.41872
		9.0	1.62449				9.0	1.35614
		10.0	1.53716				10.0	1.30243
		15.0	1.25152				15.0	1.11809
0.030	0.080	2.0	3.63368			0.120	2.0	2.57964
		3.0	2.89161				3.0	2.16816
		4.0	2.49184				4.0	1.93028
		5.0	2.23221				5.0	1.76617
		6.0	2.04499				6.0	1.64199
		7.0	1.90090				7.0	1.54284
		8.0	1.78509				8.0	1.46099
		9.0	1.68917				9.0	1.39188
		10.0	1.60797				10.0	1.33264
		15.0	1.33416				15.0	1.13103
	0.100	2.0	3.77221			0.140	2.0	2.68925
		3.0	3.01043				3.0	2.25935
		4.0	2.59373				4.0	2.00609
		5.0	2.31955				5.0	1.82916
		6.0	2.11983				6.0	1.69430
		7.0	1.96501				7.0	1.58626
		8.0	1.83999				8.0	1.49700
		9.0	1.73617				9.0	1.42174
		10.0	1.64819				10.0	1.35738
		15.0	1.35254				15.0	1.14061

$N = 2.0$

FACTOR E7   K 0.800

V	C4	LIFE	VALUE		C4	LIFE	VALUE
0.010	0.160	2.0	2.79457		0.100	2.0	3.28729
		3.0	2.34522			3.0	2.73592
		4.0	2.07607			4.0	2.41897
		5.0	1.88616			5.0	2.20092
		6.0	1.74069			6.0	2.03598
		7.0	1.62400			7.0	1.90406
		8.0	1.52769			8.0	1.79479
		9.0	1.44668			9.0	1.70212
		10.0	1.37764			10.0	1.62224
		15.0	1.14771			15.0	1.34503
0.020	0.080	2.0	2.68586		0.120	2.0	3.43940
		3.0	2.24244			3.0	2.86381
		4.0	1.99492			4.0	2.52646
		5.0	1.82857			5.0	2.29123
		6.0	1.70487			6.0	2.11183
		7.0	1.60709			7.0	1.96776
		8.0	1.52669			8.0	1.84825
		9.0	1.45882			9.0	1.74699
		10.0	1.40043			10.0	1.65988
		15.0	1.19754			15.0	1.36060
	0.100	2.0	2.82176		0.140	2.0	3.58555
		3.0	2.35956			3.0	2.98426
		4.0	2.09580			4.0	2.62568
		5.0	1.91543			5.0	2.37295
		6.0	1.77963			6.0	2.17911
		7.0	1.67140			7.0	2.02313
		8.0	1.58200			8.0	1.89382
		9.0	1.50636			9.0	1.78446
		10.0	1.44128			10.0	1.69069
		15.0	1.21654			15.0	1.37212
	0.120	2.0	2.95233		0.160	2.0	3.72597
		3.0	2.46986			3.0	3.09768
		4.0	2.18893			4.0	2.71728
		5.0	1.99403			5.0	2.44689
		6.0	1.84593			6.0	2.23878
		7.0	1.72731			7.0	2.07127
		8.0	1.62913			8.0	1.93264
		9.0	1.54606			9.0	1.81576
		10.0	1.47472			10.0	1.71592
		15.0	1.23061			15.0	1.38067
	0.140	2.0	3.07778				
		3.0	2.57373				
		4.0	2.27490				
		5.0	2.06515				
		6.0	1.90474				
		7.0	1.77592				
		8.0	1.66928				
		9.0	1.57923				
		10.0	1.50209				
		15.0	1.24104				
	0.160	2.0	3.19832				
		3.0	2.67155				
		4.0	2.35426				
		5.0	2.12950				
		6.0	1.95690				
		7.0	1.81818				
		8.0	1.70351				
		9.0	1.60693				
		10.0	1.52451				
		15.0	1.24876				
0.030	0.080	2.0	3.12896				
		3.0	2.60013				
		4.0	2.30253				
		5.0	2.10110				
		6.0	1.95045				
		7.0	1.83079				
		8.0	1.73204				
		9.0	1.64841				
		10.0	1.57627				
		15.0	1.32403				

FACTOR	T3	T4	LIFE	VALUE
E10	2.000	0.000		1.00000
		1.000		0.90909
	3.000	0.000		1.00000
		1.000		0.90909
	4.000	0.000		1.00000
		1.000		0.90909

FACTOR	T3		LIFE	VALUE
E11	2.000		2.0	0.00000
			3.0	2.47145
			4.0	1.94751
			5.0	2.98706
			6.0	2.61057
			7.0	3.22137
			8.0	2.95079
			9.0	3.36047
			10.0	3.16053
			15.0	3.57921

TABLE OF FACTORS

N = 2.0

T3	LIFE	VALUE
3.000	2.0	0.00000
	3.0	0.00000
	4.0	1.77046
	5.0	1.48677
	6.0	1.29938
	7.0	2.04377
	8.0	1.87210
	9.0	1.74051
	10.0	2.16457
	15.0	2.10034
4.000	2.0	0.00000
	3.0	0.00000
	4.0	0.00000
	5.0	1.35161
	6.0	1.18125
	7.0	1.06090
	8.0	0.97179
	9.0	1.52058
	10.0	1.43011
	15.0	1.49798

FACTOR		LIFE	VALUE
D1			0.58263

FACTOR	V	LIFE	VALUE
D2	0.010	3.0	3.29072
		4.0	2.40184
		5.0	1.89440
		6.0	1.56657
		7.0	1.33747
		8.0	1.16840
		9.0	1.03857
		10.0	0.93577
		15.0	0.63340
	0.020	3.0	3.74861
		4.0	2.72367
		5.0	2.13881
		6.0	1.76115
		7.0	1.49738
		8.0	1.30287
		9.0	1.15361
		10.0	1.03553
		15.0	0.68917
	0.030	3.0	4.34654
		4.0	3.14365
		5.0	2.45758
		6.0	2.01484
		7.0	1.70582
		8.0	1.47811
		9.0	1.30353
		10.0	1.16555
		15.0	0.76196

FACTOR	C3	LIFE	VALUE
D5	0.080		0.10099
	0.100		0.13205
	0.120		0.16189
	0.140		0.19056
	0.160		0.21810

FACTOR	K	V	C4	LIFE	VALUE
D6	0.700	0.010	0.080	2.0	1.82010
				3.0	1.31096
				4.0	1.04982
				5.0	0.88838
				6.0	0.77716
				7.0	0.69493
				8.0	0.63104
				9.0	0.57957
				10.0	0.53694
				15.0	0.39742
			0.100	2.0	1.93696
				3.0	1.41844
				4.0	1.14873
				5.0	0.97945
				6.0	0.86105
				7.0	0.77225
				8.0	0.70234
				9.0	0.64535
				10.0	0.59766
				15.0	0.43840
			0.120	2.0	2.04923
				3.0	1.51966
				4.0	1.24003
				5.0	1.06184
				6.0	0.93545
				7.0	0.83946
				8.0	0.76310
				9.0	0.70029
				10.0	0.64737
				15.0	0.46876
			0.140	2.0	2.15711
				3.0	1.61498
				4.0	1.32431
				5.0	1.13640
				6.0	1.00144
				7.0	0.89790
				8.0	0.81487
				9.0	0.74619
				10.0	0.68807
				15.0	0.49125
			0.160	2.0	2.26075
				3.0	1.70476
				4.0	1.40211
				5.0	1.20386
				6.0	1.05997
				7.0	0.94870
				8.0	0.85898
				9.0	0.78452
				10.0	0.72139
				15.0	0.50791
		0.020	0.080	2.0	2.08306
				3.0	1.49337
				4.0	1.19049
				5.0	1.00300
				6.0	0.87369
				7.0	0.77802
				8.0	0.70367
				9.0	0.64377
				10.0	0.59419
				15.0	0.43241
			0.100	2.0	2.21680
				3.0	1.61581
				4.0	1.30265
				5.0	1.10581
				6.0	0.96800
				7.0	0.86458
				8.0	0.78317
				9.0	0.71684
				10.0	0.66138
				15.0	0.47700

CONSTRUCTION EQUIPMENT POLICY

$N = 2.0$

FACTOR D6   K 0.700

V	C4	LIFE	VALUE	K	V	C4	LIFE	VALUE
0.020	0.120	2.0	2.34530			0.160	2.0	3.01424
		3.0	1.73111				3.0	2.25173
		4.0	1.40619				4.0	1.83516
		5.0	1.19884				5.0	1.56176
		6.0	1.05164				6.0	1.36327
		7.0	0.93983				7.0	1.20998
		8.0	0.85092				8.0	1.08668
		9.0	0.77786				9.0	0.98467
		10.0	0.71639				10.0	0.89853
		15.0	0.51003				15.0	0.61100
	0.140	2.0	2.46876	0.800	0.010	0.080	2.0	1.39434
		3.0	1.83970				3.0	1.04730
		4.0	1.50176				4.0	0.86605
		5.0	1.28302				5.0	0.75168
		6.0	1.12583				6.0	0.67119
		7.0	1.00525				7.0	0.61039
		8.0	0.90865				8.0	0.56216
		9.0	0.82884				9.0	0.52253
		10.0	0.76142				10.0	0.48909
		15.0	0.53450				15.0	0.37484
	0.160	2.0	2.58738			0.100	2.0	1.52789
		3.0	1.94197				3.0	1.17013
		4.0	1.58999				4.0	0.97908
		5.0	1.35918				5.0	0.85575
		6.0	1.19162				6.0	0.76706
		7.0	1.06213				7.0	0.69875
		8.0	0.95784				8.0	0.64364
		9.0	0.87142				9.0	0.59770
		10.0	0.79830				10.0	0.55848
		15.0	0.55262				15.0	0.42167
0.030	0.080	2.0	2.42672			0.120	2.0	1.65621
		3.0	1.73158				3.0	1.28581
		4.0	1.37406				4.0	1.08342
		5.0	1.15249				5.0	0.94992
		6.0	0.99955				6.0	0.85209
		7.0	0.88632				7.0	0.77556
		8.0	0.79832				8.0	0.71307
		9.0	0.72744				9.0	0.66049
		10.0	0.66879				10.0	0.61530
		15.0	0.47809				15.0	0.45636
	0.100	2.0	2.58252			0.140	2.0	1.77950
		3.0	1.87354				3.0	1.39476
		4.0	1.50352				4.0	1.17975
		5.0	1.27063				5.0	1.03513
		6.0	1.10744				6.0	0.92750
		7.0	0.98493				7.0	0.84235
		8.0	0.88851				8.0	0.77224
		9.0	0.81000				9.0	0.71294
		10.0	0.74442				10.0	0.66181
		15.0	0.52738				15.0	0.48206
	0.120	2.0	2.73222			0.160	2.0	1.89795
		3.0	2.00724				3.0	1.49736
		4.0	1.62302				4.0	1.26866
		5.0	1.37752				5.0	1.11223
		6.0	1.20313				6.0	0.99439
		7.0	1.07066				7.0	0.90040
		8.0	0.96537				8.0	0.82266
		9.0	0.87896				9.0	0.75675
		10.0	0.80633				10.0	0.69989
		15.0	0.56390				15.0	0.50110
	0.140	2.0	2.87605		0.020	0.080	2.0	1.59579
		3.0	2.13315				3.0	1.19303
		4.0	1.73333				4.0	0.98209
		5.0	1.47424				5.0	0.84866
		6.0	1.28800				6.0	0.75455
		7.0	1.14519				7.0	0.68337
		8.0	1.03087				8.0	0.62686
		9.0	0.93655				9.0	0.58041
		10.0	0.85703				10.0	0.54123
		15.0	0.59095				15.0	0.40784

TABLE OF FACTORS   283

$N = 2.0$

V	C4	LIFE	VALUE		C4	LIFE	VALUE
	0.100	2.0	1.74864		0.140	2.0	2.37258
		3.0	1.33295			3.0	1.84226
		4.0	1.11027			4.0	1.54412
		5.0	0.96616			5.0	1.34286
		6.0	0.86233			6.0	1.19290
		7.0	0.78229			7.0	1.07434
		8.0	0.71771			8.0	0.97694
		9.0	0.66391			9.0	0.89483
		10.0	0.61802			10.0	0.82432
		15.0	0.45879			15.0	0.57991
	0.120	2.0	1.89549		0.160	2.0	2.53051
		3.0	1.46473			3.0	1.97778
		4.0	1.22860			4.0	1.66050
		5.0	1.07248			5.0	1.44288
		6.0	0.95792			6.0	1.27893
		7.0	0.86830			7.0	1.14839
		8.0	0.79514			8.0	1.04073
		9.0	0.73366			9.0	0.94982
		10.0	0.68089			10.0	0.87175
		15.0	0.49654			15.0	0.60281
	0.140	2.0	2.03659				
		3.0	1.58883				
		4.0	1.33783				
		5.0	1.16868	FACTOR			VALUE
		6.0	1.04271				
		7.0	0.94306	T1			0.09445
		8.0	0.86112				
		9.0	0.79191				
		10.0	0.73236				
		15.0	0.52451	FACTOR	V	LIFE	VALUE
	0.160	2.0	2.17216				
		3.0	1.70571	T2	0.010	2.0	0.96008
		4.0	1.43866			3.0	0.94697
		5.0	1.25572			4.0	0.93450
		6.0	1.11790			5.0	0.77806
		7.0	1.00806			6.0	0.67439
		8.0	0.91734			7.0	0.60088
		9.0	0.84058			8.0	0.54621
		10.0	0.77451			9.0	0.50410
		15.0	0.54522			10.0	0.47077
						15.0	0.37427
0.030	0.080	2.0	1.85906				
		3.0	1.38332		0.020	2.0	1.09879
		4.0	1.13353			3.0	1.07874
		5.0	0.97515			4.0	1.05972
		6.0	0.86324			5.0	0.87844
		7.0	0.77849			6.0	0.75816
		8.0	0.71117			7.0	0.67273
		9.0	0.65584			8.0	0.60907
		10.0	0.60919			9.0	0.55993
		15.0	0.45092			10.0	0.52095
						15.0	0.40723
	0.100	2.0	2.03712				
		3.0	1.54557		0.030	2.0	1.28006
		4.0	1.28147			3.0	1.25081
		5.0	1.11016			4.0	1.22313
		6.0	0.98655			5.0	1.00937
		7.0	0.89119			6.0	0.86737
		8.0	0.81425			7.0	0.76637
		9.0	0.75019			8.0	0.69100
		10.0	0.69562			9.0	0.63270
		15.0	0.50725			10.0	0.58636
						15.0	0.45024
	0.120	2.0	2.20821				
		3.0	1.69837				
		4.0	1.41805				
		5.0	1.23232				
		6.0	1.09591				
		7.0	0.98916				
		8.0	0.90209				
		9.0	0.82900				
		10.0	0.76638				
		15.0	0.54899				

$N = 2.0$

FACTOR	C3					VALUE
G1	0.080					0.52692
	0.100					0.49214
	0.120					0.45873
	0.140					0.42663
	0.160					0.39578

FACTOR	K	V	C4	LIFE	VALUE
G2	0.700	0.010	0.080	3.0	1.44690
				4.0	0.89756
				5.0	0.60135
				6.0	0.42207
				7.0	0.30540
				8.0	0.22567
				9.0	0.16928
				10.0	0.12836
				15.0	0.03474
			0.100	3.0	1.35694
				4.0	0.81972
				5.0	0.53403
				6.0	0.36389
				7.0	0.25513
				8.0	0.18227
				9.0	0.13183
				10.0	0.09607
				15.0	0.01946
			0.120	3.0	1.27222
				4.0	0.74786
				5.0	0.47312
				6.0	0.31228
				7.0	0.21143
				8.0	0.14529
				9.0	0.10055
				10.0	0.06963
				15.0	0.00814
			0.140	3.0	1.19243
				4.0	0.68153
				5.0	0.41800
				6.0	0.26651
				7.0	0.17344
				8.0	0.11378
				9.0	0.07443
				10.0	0.04798
				15.0	-0.00024
			0.160	3.0	1.11729
				4.0	0.62029
				5.0	0.36813
				6.0	0.22591
				7.0	0.14042
				8.0	0.08693
				9.0	0.05260
				10.0	0.03026
				15.0	-0.00646
		0.020	0.080	3.0	1.64823
				4.0	1.01783
				5.0	0.67894
				6.0	0.47450
				7.0	0.34191
				8.0	0.25164
				9.0	0.18803
				10.0	0.14204
				15.0	0.03780

TABLE OF FACTORS

$N = 2.0$

V	C4	LIFE	VALUE	K	V	C4	LIFE	VALUE
	0.100	3.0	1.54576			0.160	3.0	1.47578
		4.0	0.92956				4.0	0.81187
		5.0	0.60293				5.0	0.47757
		6.0	0.40908				6.0	0.29055
		7.0	0.28564				7.0	0.17909
		8.0	0.20325				8.0	0.10997
		9.0	0.14643				9.0	0.06603
		10.0	0.10631				10.0	0.03769
		15.0	0.02117				15.0	-0.00777
	0.120	3.0	1.44925	0.800	0.010	0.080	3.0	1.66758
		4.0	0.84807				4.0	1.04220
		5.0	0.53416				5.0	0.70241
		6.0	0.35106				6.0	0.49558
		7.0	0.23671				7.0	0.36036
		8.0	0.16201				8.0	0.26760
		9.0	0.11169				9.0	0.20175
		10.0	0.07705				10.0	0.15381
		15.0	0.00886				15.0	0.04316
	0.140	3.0	1.35836			0.100	3.0	1.56477
		4.0	0.77285				4.0	0.95324
		5.0	0.47193				5.0	0.62548
		6.0	0.29961				6.0	0.42908
		7.0	0.19418				7.0	0.30292
		8.0	0.12687				8.0	0.21800
		9.0	0.08267				9.0	0.15895
		10.0	0.05309				10.0	0.11690
		15.0	-0.00027				15.0	0.02570
	0.160	3.0	1.27276			0.120	3.0	1.46795
		4.0	0.70341				4.0	0.87112
		5.0	0.41562				5.0	0.55586
		6.0	0.25397				6.0	0.37010
		7.0	0.15721				7.0	0.25298
		8.0	0.09693				8.0	0.17574
		9.0	0.05843				9.0	0.12321
		10.0	0.03348				10.0	0.08669
		15.0	-0.00703				15.0	0.01276
0.030	0.080	3.0	1.91114			0.140	3.0	1.37676
		4.0	1.17478				4.0	0.79531
		5.0	0.78013				5.0	0.49287
		6.0	0.54285				6.0	0.31779
		7.0	0.38951				7.0	0.20956
		8.0	0.28549				8.0	0.13973
		9.0	0.21246				9.0	0.09335
		10.0	0.15988				10.0	0.06195
		15.0	0.04179				15.0	0.00318
	0.100	3.0	1.79231			0.160	3.0	1.29088
		4.0	1.07289				4.0	0.72532
		5.0	0.69279				5.0	0.43587
		6.0	0.46801				6.0	0.27140
		7.0	0.32540				7.0	0.17181
		8.0	0.23059				8.0	0.10904
		9.0	0.16546				9.0	0.06841
		10.0	0.11966				10.0	0.04169
		15.0	0.02341				15.0	-0.00392
	0.120	3.0	1.68041		0.020	0.080	3.0	1.89962
		4.0	0.97884				4.0	1.18186
		5.0	0.61377				5.0	0.79304
		6.0	0.40163				6.0	0.55714
		7.0	0.26967				7.0	0.40345
		8.0	0.18381				8.0	0.29840
		9.0	0.12621				9.0	0.22410
		10.0	0.08672				10.0	0.17021
		15.0	0.00979				15.0	0.04696
	0.140	3.0	1.57503					
		4.0	0.89202					
		5.0	0.54227					
		6.0	0.34276					
		7.0	0.22121					
		8.0	0.14394					
		9.0	0.09341					
		10.0	0.05976					
		15.0	-0.00029					

CONSTRUCTION EQUIPMENT POLICY

$N = 2.0$

FACTOR	G2	K	0.800					
V	C4	LIFE	VALUE			C4	LIFE	VALUE
0.020	0.100	3.0	1.78250			0.160	3.0	1.70507
		4.0	1.08097				4.0	0.94934
		5.0	0.70617				5.0	0.56545
		6.0	0.48238				6.0	0.34905
		7.0	0.33914				7.0	0.21913
		8.0	0.24309				8.0	0.13794
		9.0	0.17656				9.0	0.08586
		10.0	0.12937				10.0	0.05193
		15.0	0.02796				15.0	-0.00472
	0.120	3.0	1.67221					
		4.0	0.98784					
		5.0	0.62757					
		6.0	0.41607	FACTOR	V		LIFE	VALUE
		7.0	0.28322	IC11	0.010			0.84314
		8.0	0.19597					
		9.0	0.13686		0.020			0.86017
		10.0	0.09593					
		15.0	0.01389		0.030			0.87755
	0.140	3.0	1.56833					
		4.0	0.90187					
		5.0	0.55646					
		6.0	0.35726					
		7.0	0.23462	FACTOR	C3		LIFE	VALUE
		8.0	0.15581	IC12	0.080			0.70425
		9.0	0.10369					
		10.0	0.06855		0.100			0.67664
		15.0	0.00346					
	0.160	3.0	1.47051		0.120			0.65011
		4.0	0.82251					
		5.0	0.49210		0.140			0.62461
		6.0	0.30511					
		7.0	0.19236		0.160			0.60012
		8.0	0.12158					
		9.0	0.07599					
		10.0	0.04614					
		15.0	-0.00427					

FACTOR				FACTOR	K	V	C4		
0.030	0.080	3.0	2.20262	IC2	0.700	0.010	0.080		
		4.0	1.36409					3.0	1.34607
		5.0	0.91123					4.0	1.06069
		6.0	0.63739					5.0	0.87753
		7.0	0.45961					6.0	0.74688
		8.0	0.33854					7.0	0.64728
		9.0	0.25322					8.0	0.56791
		10.0	0.19158					9.0	0.50267
		15.0	0.05192					10.0	0.44783
								15.0	0.26591
	0.100	3.0	2.06682						
		4.0	1.24765						
		5.0	0.81142				0.100		
		6.0	0.55186					3.0	1.43603
		7.0	0.38634					4.0	1.13854
		8.0	0.27579					5.0	0.94485
		9.0	0.19951					6.0	0.80506
		10.0	0.14561					7.0	0.69755
		15.0	0.03091					8.0	0.61131
								9.0	0.54012
	0.120	3.0	1.93893					10.0	0.48012
		4.0	1.14016					15.0	0.28119
		5.0	0.72111						
		6.0	0.47601				0.120		
		7.0	0.32265					3.0	1.52075
		8.0	0.22232					4.0	1.21040
		9.0	0.15464					5.0	1.00577
		10.0	0.10797					6.0	0.85667
		15.0	0.01535					7.0	0.74124
								8.0	0.64829
	0.140	3.0	1.81849					9.0	0.57140
		4.0	1.04094					10.0	0.50656
		5.0	0.63939					15.0	0.29251
		6.0	0.40873						
		7.0	0.26727				0.140		
		8.0	0.17676					3.0	1.60054
		9.0	0.11717					4.0	1.27673
		10.0	0.07716					5.0	1.06089
		15.0	0.00382					6.0	0.90244
								7.0	0.77923
								8.0	0.67980
								9.0	0.59752
								10.0	0.52821
								15.0	0.30089

TABLE OF FACTORS 287

$N = 2.0$

V	C4	LIFE	VALUE	K	V	C4	LIFE	VALUE
	0.160					0.100		
		3.0	1.67568				3.0	2.13289
		4.0	1.33797				4.0	1.71618
		5.0	1.11076				5.0	1.44200
		6.0	0.94304				6.0	1.24228
		7.0	0.81226				7.0	1.08746
		8.0	0.70666				8.0	0.96244
		9.0	0.61935				9.0	0.85862
		10.0	0.54593				10.0	0.77064
		15.0	0.30711				15.0	0.47499
0.020	0.080					0.120		
		3.0	1.63366				3.0	2.24479
		4.0	1.29877				4.0	1.81023
		5.0	1.08250				5.0	1.52102
		6.0	0.92734				6.0	1.30866
		7.0	0.80845				7.0	1.14319
		8.0	0.71328				8.0	1.00922
		9.0	0.63472				9.0	0.89787
		10.0	0.56843				10.0	0.80358
		15.0	0.34653				15.0	0.48861
	0.100					0.140		
		3.0	1.73613				3.0	2.35018
		4.0	1.38704				4.0	1.89705
		5.0	1.15850				5.0	1.59252
		6.0	0.99275				6.0	1.36753
		7.0	0.86473				7.0	1.19164
		8.0	0.76167				8.0	1.04909
		9.0	0.67631				9.0	0.93066
		10.0	0.60416				10.0	0.83054
		15.0	0.36315				15.0	0.49869
	0.120					0.160		
		3.0	1.83264				3.0	2.44943
		4.0	1.46853				4.0	1.97720
		5.0	1.22728				5.0	1.65722
		6.0	1.05077				6.0	1.41974
		7.0	0.91365				7.0	1.23377
		8.0	0.80290				8.0	1.08306
		9.0	0.71105				9.0	0.95805
		10.0	0.63342				10.0	0.85261
		15.0	0.37547				15.0	0.50617
	0.140			0.800	0.010	0.080		
		3.0	1.92353				3.0	1.12539
		4.0	1.54375				4.0	0.91605
		5.0	1.28951				5.0	0.77647
		6.0	1.10223				6.0	0.67336
		7.0	0.95618				7.0	0.59231
		8.0	0.83804				8.0	0.52598
		9.0	0.74007				9.0	0.47020
		10.0	0.65738				10.0	0.42238
		15.0	0.38459				15.0	0.25749
	0.160					0.100		
		3.0	2.00913				3.0	1.22820
		4.0	1.61319				4.0	1.00502
		5.0	1.34581				5.0	0.85341
		6.0	1.14787				6.0	0.73987
		7.0	0.99316				7.0	0.64976
		8.0	0.86799				8.0	0.57558
		9.0	0.76431				9.0	0.51300
		10.0	0.67699				10.0	0.45929
		15.0	0.39135				15.0	0.27495
0.030	0.080					0.120		
		3.0	2.01406				3.0	1.32502
		4.0	1.61429				4.0	1.08714
		5.0	1.35466				5.0	0.92303
		6.0	1.16744				6.0	0.79885
		7.0	1.02335				7.0	0.69970
		8.0	0.90754				8.0	0.61784
		9.0	0.81162				9.0	0.54874
		10.0	0.73042				10.0	0.48950
		15.0	0.45661				15.0	0.28789

$N = 2.0$

FACTOR IC2 K 0.800

V	C4	LIFE	VALUE	V	C4	LIFE	VALUE
0.010	0.140			0.030	0.080		
		3.0	1.41621			3.0	1.72258
		4.0	1.16295			4.0	1.42498
		5.0	0.98602			5.0	1.22355
		6.0	0.85116			6.0	1.07290
		7.0	0.74312			7.0	0.95324
		8.0	0.65386			8.0	0.85449
		9.0	0.57860			9.0	0.77086
		10.0	0.51424			10.0	0.69872
		15.0	0.29747			15.0	0.44648
	0.160				0.100		
		3.0	1.50208			3.0	1.85838
		4.0	1.23293			4.0	1.54142
		5.0	1.04301			5.0	1.32337
		6.0	0.89755			6.0	1.15843
		7.0	0.78086			7.0	1.02651
		8.0	0.68455			8.0	0.91724
		9.0	0.60354			9.0	0.82457
		10.0	0.53450			10.0	0.74469
		15.0	0.30457			15.0	0.46748
0.020	0.080				0.120		
		3.0	1.38227			3.0	1.98627
		4.0	1.13475			4.0	1.64891
		5.0	0.96840			5.0	1.41368
		6.0	0.84469			6.0	1.23428
		7.0	0.74691			7.0	1.09021
		8.0	0.66652			8.0	0.97070
		9.0	0.59865			9.0	0.86944
		10.0	0.54026			10.0	0.78233
		15.0	0.33736			15.0	0.48305
	0.100				0.140		
		3.0	1.49939			3.0	2.10671
		4.0	1.23563			4.0	1.74814
		5.0	1.05526			5.0	1.49540
		6.0	0.91945			6.0	1.30156
		7.0	0.81123			7.0	1.14558
		8.0	0.72182			8.0	1.01627
		9.0	0.64618			9.0	0.90691
		10.0	0.58110			10.0	0.81314
		15.0	0.35636			15.0	0.49457
	0.120				0.160		
		3.0	1.60968			3.0	2.22014
		4.0	1.32876			4.0	1.83973
		5.0	1.13386			5.0	1.56934
		6.0	0.98576			6.0	1.36124
		7.0	0.86714			7.0	1.19372
		8.0	0.76895			8.0	1.05509
		9.0	0.68589			9.0	0.93821
		10.0	0.61454			10.0	0.83837
		15.0	0.37044			15.0	0.50311
	0.140						
		3.0	1.71356				
		4.0	1.41473				
		5.0	1.20498				
		6.0	1.04457				
		7.0	0.91575				
		8.0	0.80911				
		9.0	0.71905				
		10.0	0.64192				
		15.0	0.38087				
	0.160						
		3.0	1.81138				
		4.0	1.49409				
		5.0	1.26933				
		6.0	1.09673				
		7.0	0.95801				
		8.0	0.84333				
		9.0	0.74675				
		10.0	0.66433				
		15.0	0.38859				

Appendix 4
Chart of Accounts

Cost acct. no.	Cost subacct. no.	Cost type	Item
.10			Total profit
.11			Profit per operating hour (mile)
.12			Cost per operating hour (mile)
.13			Profit per unit production
.14			Total operating hours (miles)
.15			Total production
.16			Production per operating hour
.17			% utilization
.18			% availability
.19			Total costs
.20			Total revenue
.21			Outside rental
.22			Internal rental
.23			Other
.30			Total fixed costs
.31			Depreciation (straight line)
.32	00		Total interest, insurance, and taxes
	01		Interest on investment
	02		Loan interest

Cost acct. no.	Cost subacct. no.	Cost type	Item
	03		Other interest
	04		PL & PD insurance
	05		Collision insurance
	06		Fire insurance
	07		Comprehensive insurance
	08		Other insurance
	09		Personal property tax
	10		Inventory tax
	11		Use tax
	12		Other taxes
.33	00		Total storage and security
	01		Off-season storage
	02		Job storage
	03		Off-season security
	04		Job security
	05		Other
.34	00		Total fees, licenses, and fines
	01		Use permits
	02		Annual license fee
	03		Special fees
	04		Fines for speeding
	05		Fines for overload
	06		Other fines
	07		Other
.35	00		Total moving costs
	01		Transportation
	02		Setup
	03		Dismantle
	04		Other
.36	00		Total overhead costs
	01		General expense
	02		Job overhead
	03		Other
.37			Total, other fixed costs
.40			Total variable costs
.41	00	0	Total scheduled maintenance
	01	1	Inspection
	02	1	Adjustment
	03	–	Crankcase lube
	04	–	Grease
	05	–	Hydraulic
	06	–	Coolant
	07	–	Oil filter
	08	–	Air filter
	09	–	Other
.42	00	0	Total field repairs
	01	0	Field repair cost per operating hour (mile)
	10	0	Total attachments cost

CHART OF ACCOUNTS

Cost acct. no.	Cost subacct. no.	Cost type	Item
	11	–	Electric (generator, winch, etc.)
	12	–	Hydraulic (pcu, ripper, etc.)
	13	–	Mechanical (pcu, winch, blade, etc.)
	14	–	Other
	20	–	Total chassis cost
	21	–	Body
	22	–	Frame
	23	–	Front suspension
	24	–	Rear suspension
	25	–	Other
	30	0	Total drive train cost
	31	–	Clutch
	32	–	Differential
	33	–	Drive shaft
	34	–	Final drive
	35	–	Transmission, power shift
	36	–	Transmission, gear shift
	37	–	Other
	40	0	Total engine cost
	41	–	Bearings
	42	–	Cam
	43	–	Crank
	44	–	Cylinders
	45	–	Pistons
	46	–	Valves
	47	–	Cooling system
	48	–	Other
	50	0	Total brakes cost
	51	–	Linings
	52	–	Master cylinder
	53	–	Wheel cylinders
	54	–	Hydraulic lines
	55	–	Drums
	56	–	Retarder
	57	–	Other
	60	0	Total running gear
	61	–	Wheels
	62	–	Running chains (mud or rock)
	63	–	Wheel shoes
	64	–	Other
	70	0	Total electrical system
	71	–	Spark plugs
	72	–	Distributor
	73	–	Other ignition
	74	–	Generator/alternator
	75	–	Voltage regulator
	76	–	Battery
	77	–	Other

Cost acct. no.	Cost subacct. no.	Cost type	Item
	80	0	Total fuel-system cost
	81	–	Carburetor
	82	–	Pump
	83	–	Injectors
	84	–	Tank
	85	–	Lines
	86	–	Other
	90	0	Total steering cost
	91	–	Linkage
	92	–	Pump
	93	–	Hydraulic lines
	94	–	Other
.43	00		Total fuel cost
	01		Fuel cost per operating hour (mile)
	02		Diesel fuel
	03		Gasoline
	04		Liquefied petroleum gas
	05		Additives
	06		Other
.44	00	0	Total running gear
	01	0	Cost per operating hour (miles)
	01	2	Tires, wear
	02	–	Tires, repair
	03	1	Tires, change
	04	–	Track shoes
	05	–	Track rails
	06	–	Track pins
	07	–	Idlers and rollers
	08	–	Drive sprocket
	09	–	Other track
.45	00		Total operating labor
	01		Cost per operating hour (mile)
	02		Operator, straight-time cost
	03		Operator, premium-time cost
	04		Assistant, straight-time cost
	05		Assistant, premium-time cost
	06		Other, straight-time cost
	07		Other, premium-time cost
.46	00		Total cost, field supervision
	01		Supervision cost per operating hour
	02		Supervision, straight time
	03		Supervision, premium time
.47	00	0	Total cost, major repairs
			Subitems same as .42
.48	00	0	Total cost, overhauls
			Subitems same as .42
.49			Total, other variable costs

CHART OF ACCOUNTS

Time acct. no.	Time subacct. no.	Time type	Item
.50			Total shift hours
.51	00		Total idle time
	01		Weather on job
	02		Seasonal shut down
	03		Strike or labor delay
	04		In transit
	05		Setup
	06		Dismantle
	07		Other
	08		Other
.52			Total scheduled hours
.53	00		Total downtime
	01	–	Accident
	02	–	Breakdown
	03	–	Scheduled repairs
	04		Preventive maintenance
	05		Warmup
	06		Shut down
.54			Total operating hours
.55			Other time
.60			Total units produced
.61			Other production
.70			Other costs

Cost type		Time type	
0	Combined	1	Waiting parts
1	Labor	2	Waiting repair
2	Material	3	In repair
3	Outside contract		
4	Other		

Appendix 5
Equipment Classification Code

Group code
Type code

Group 0: Passenger-carrying vehicles and emergency vehicles

01	Bus
02	Sedan
03	Carryall
04	Ambulance
05	Fire trucks
06	Salvage (or crash) trucks and wreckers
07	Rescue trucks
08	Other
09	Vacant

Group 1: Highway trucks and trailers

10	Pickup trucks
11	Light trucks ($\frac{1}{4}$–2 tons)
12	Medium trucks (2–5 tons)
13	Heavy trucks (> 5 tons)
14	Tank trucks

EQUIPMENT CLASSIFICATION CODE

	15	Utility trailers
	16	Semitrailers
	17	Full trailers
	18	Other
	19	Vacant

Group 2: Aggregate production and paving

	20	Bins
	21	Mixers
	22	Crushers
	23	Speaders and finishers
	24	Pavers
	25	Screening and classifying
	26	Dehydrators
	27	Concrete placers (pumps, air, gunite, etc.)
	28	Other
	29	Vacant

Group 3: Drilling, pile driving, and compaction

	30	Air compressors
	31	Drills
	32	Pile hammers
	33	Pile leads and accessories
	34	Steel wheel compactors and rollers
	35	Rubber-tired compactors and rollers
	36	Vibrating compactors
	37	Sheepsfoot rollers
	38	Other
	39	Vacant

Group 4: Excavating and loading

	40	Excavators (shovels, backhoes, bucket wheels, elevating graders)
	41	Ditchers
	42	Rippers and rooters
	43	Crawler loaders
	44	Wheel loaders
	45	Belt loaders
	46	Other
	47	Vacant
	48	Vacant
	49	Vacant

Group 5: Grading and hauling

	50	Motor graders
	51	Subgraders and fine graders
	52	Crawler tractors

	53	Wheel tractors
	54	Towed scrapers
	55	Wheel tractor-scrapers
	56	Off-highway trucks
	57	Off-highway wagons
	58	Off-highway trailers
	59	Other

Group 6: Miscellaneous construction and maintenance

	60	Electric generators
	61	Welders
	62	Pumps
	63	Plows (snow, cable, disk, etc.)
	64	Mowers (reel, rotary, sickle-bar, hammer-knife, etc.)
	65	Other
	66	Vacant
	67	Vacant
	68	Vacant
	69	Vacant

Group 7: Railway, mining, and marine

	70	Locomotives (including inspection and service cars)
	71	Cars (box, flat, dump, mine, tank, etc.)
	72	Other railroad equipment (cranes, tampers, etc.)
	73	Tunnel shields
	74	Moles and accessory equipment
	75	Other mining equipment
	76	Boats (tug, work, etc.)
	77	Barges (scow, dump, work, crane, etc.)
	78	Dredges
	79	Other marine equipment

Group 8: Weight handling

	80	Crawler cranes
	81	Truck cranes
	82	Cruiser cranes
	83	Tower cranes
	84	Straddle cranes
	85	Winches
	86	Cableways
	87	Other
	88	Vacant
	89	Vacant

EQUIPMENT CLASSIFICATION CODE

Group 9: Materials handling

90	Warehouse tractors
91	Warehouse cranes
92	Forklift trucks
93	Straddle carriers
94	Logging (skidders, arches, etc.)
95	Conveyors (screw, belt, bucket elevator, etc.)
96	Air-transfer equipment
97	Other
98	Vacant
99	Vacant

Index

Abuse of equipment, 215
Accidents, cost of, 222–224
Alternatives, differences between, 17
ASCE survey, 11
Associated Equipment Distributors (AED), 132–134
Assumptions in tables, 78
Automatic lubrication systems, 214
Availability, 105–109

Bank loans, 120
Blue Book Rental Rates, 132, 135
Brakes, 231
Bulletin F, 35
Business ratios, 140

Cannibalization, 215
Consumers Price Index (CPI), 41–44
Cost curves:
 capital, 64
 depreciation, 66–68
 discrete, 65
 interest, 68
 loss on sale, 69
 maintenance and operating, 63, 107–114
 overhaul, 65
 variable costs versus operating hours, figure, 113

Costs:
 basic accounts, 104
 chart of accounts, 103–105
 clearing accounts, 105
 collection of, 102
 difficulty in combining, 100
 historical, 101
 nature of, 99
 obtaining, at constant demand, 106–113
 records, 97–99
Custody of equipment, 189–191

Depreciation:
 half-year convention, 23
 methods, 29–34
 as tax incentive, 29
de Stwolinski, L. W., 225, 240
Discount factors, 141–144
Discounted cash flow analysis, 141–150
Douglas, James, $33n.$, $44n.$
Dryden, L. S., 209
Dudick, T. S., $137n.$
Dynamometer, 209–211

Ease of maintenance, 216–218
Economic life, 25, 49, 89
 by computer, 93–95
 by intuition, 56
 by mathematical model, 58–73
 by maximizing profits, 54–56

299

Economic life:
  by minimizing costs, 52–54
  by tables, 75–81, 88–90
Engine families:
  Caterpillar, 159
  Corps of Engineers, 161–165
  Cummins, 158
  General Motors, 156, 160
  International Harvester, 161
Engineering and Grading Contractors Association of California (EGCA), 132
Engineering economy, 15–18
Equipment classification codes, 177, 180–182
Equipment Guide Book Company, 86n., 132
Equipment life:
  importance of, 47
  solution by tables, 75–81
Exponential curves:
  determination of, 82–87
  use of, 60

Factors:
  discount, 141–144
  economic, 45
  table, 76
Fenders, 231
Financing equipment, 117
Forms for records, 178–180, 195, 205, 222

Grant, E. L., 15, 29
Grimaldi, J. V., 226
Growth of machines:
  performance, 7

Growth of machines:
  size, 6
  technology, 6–10

Ignition analyzer, 211–214
Inflation, 25, 39–44
Inspection of equipment, 206
Interest:
  add-on, 122–124
  compound, 18–24
  discount factors, 142
  effective rate, 20
  force of, 20
  nominal rate, 19
  present worth, 20–22
  simple, 121
Inventory:
  annual, 190
  national equipment, 4
  spare parts, 11, 169
Inventory data card, 178
Investment credit, 38
Ireson, W. G., 15

Koster, F. D., 152, 169–171

Leasing equipment:
  advantages, 124
  conditional sales, 131
  master leases, 129
  payments, 125–129
  tax implications, 130–132
  types of leases, 129
Lewis, B. T., 195n., 202
Life:
  economic, 25, 49
  physical, 48
  profit, 48
  replacement, 50

Life:
  useful service, 34–37
  (*See also* Economic life; Replacement life)
Lubricating oil analysis, 207–209

MacNabb, R. R., 118
Maintenance:
  preventive, 196–199, 203–206
  records, 195
  scheduled, 193, 199–201
  unscheduled, 201
Management reports, 112–114
Marron, J. P., 195*n*., 202
Mathematical model for equipment life, 58–73

Noise suppression, 232–234
Norton, P. T., Jr., 29
Numbering equipment, 176, 180–182

Obsolescence, 44
Occupational Safety and Health Act (OSHA), 220–222, 234, 235, 238, 241
Operating hours versus scheduled hours, figure, 112
Ownership of equipment, 5, 199
  (*See also* Inventory)

Pilferage, 187–189
Policy:
  definition, 2

Policy:
  elements of, 3, 14
  purpose of, 1–3
Present worth, 20–22
Preventive maintenance, 196–199, 203–206
Price indexes, 41–44
Purchase of equipment:
  by bank loan, 120–124
  by cash, 119

Recapture of income, 39
Records, maintenance, 195
Registration of equipment, 176–177
Renting equipment, 132–139
  general conditions of, 137
  with option to buy, 139
  rates for, 132–137
Replacement life, 50, 72, 90–92
  by computer, 93–95
  by intuition, 56
  by mathematical model, 58–73
  by maximizing profits, 56
  by minimizing costs, 54
  by tables, 75–81, 90–92
Return on investment, 21–23
Revenue:
  curves, 62
  gain on sale, 69–70
  investment credit, 69
Reverse alarms, 232
Roll over protective structures (ROPS), 227, 229–231

Safety:
  BLS survey, 225

Safety:
  devices: brakes, 231
    crane indicators, 234
    fenders, 231
    reverse alarms, 232
    ROPS, 229–231
    seat belts, 231
  elements of, 224
  equipment, 227–232,
      239–241
  personal gear, 227
  personnel, 226
  policy, 234–238
  training, 238
Salvage value, 29, 37
Scheduled maintenance, 193,
    199–201
Schuster, Ray, 118
Seat belts, 231
Security, 185–189
Separable decisions, 16
Simonds, R. H., 226
Spare parts:
  mortality, 162
  support, 169–171
Speed of travel, 9
Standardization:
  benefits, 168–171
  disadvantages, 171
  economic advantages,
      165–168
  engine families, 156–165
  examples of, 172–174
  levels of, 153
Stewart, R. S., 107
Storage:
  effect of climate on, 184

Storage:
  long-term, 183
  short-term, 183
Sunk costs, 26
Surveys of equipment policy:
  ASCE, 11
  Stanford University, 13

Tables for solution:
  assumptions, 78
  range of values, 79
Technology, influence of, 6–8
Theft, 187–189
Time value of money, 18–24
Transfer of equipment, 190

Useful service life, 34–37
Utilization, 105

Vandalism, 185–187
Variables:
  choosing values, 82–87
  for solution, equipment life,
      77–81

Wear detectors, 214
Wheel tractor-scrapers:
  capacity, figure, 7
  horsepower, figure, 8
  maximum speed, figure, 8
Williams-Steiger Act (see Occupational Safety and Health Act)